SPACEWEAR

SPACEWEAR

Weightlessness and the Final Frontier of Fashion

BARBARA BROWNIE

BLOOMSBURY VISUAL ARTS
LONDON • NEW YORK • OXFORD • NEW DELHI • SYDNEY

BLOOMSBURY VISUAL ARTS
Bloomsbury Publishing Plc
50 Bedford Square, London, WC1B 3DP, UK
1385 Broadway, New York, NY 10018, USA

BLOOMSBURY, BLOOMSBURY VISUAL ARTS and the Diana logo are trademarks of Bloomsbury Publishing Plc

First published in Great Britain 2019

Cover design by Adriana Brioso
Cover images © Henrik Sorensen/Getty Images & Jake Weirick/Unsplash

A catalogue record for this book is available from the British Library.

A catalog record for this book is available from the Library of Congress.

ISBN: HB: 978-1-350-00032-2
ePDF: 978-1-350-00033-9
eBook: 978-1-350-00034-6

Typeset by Newgen KnowledgeWorks Pvt. Ltd., Chennai, India

To find out more about our authors and books visit www.bloomsbury.com and sign up for our newsletters.

CONTENTS

List of illustrations vi
Acknowledgments viii

Introduction: The final frontier of fashion 1

1 Space style 11
The first Space Race 11
The commercial space age 19
The art and design legacy of the Space Age 26

2 Weightlessness on (and above) the catwalk 41
Suspension and rigged displays 42
Floating fashion 51
Free-falling fashion models 56

3 Commercial spacewear 69
Form, function, and fantasy 70
Suiting the space tourist 78
Off-the-shelf and out-of-this-world 83

4 The clothed body in microgravity 93
"One small step": Movement in microgravity 94
Posture, shape, and the weightless body 100
Drape in weightlessness 107

Conclusion: Designing post-gravity fashion 117

Glossary 131
Notes 135
Bibliography 141
Index 161

ILLUSTRATIONS

1.1 Astronaut Edward H. White II floats above the Earth on the first spacewalk 14
1.2 Pierre Cardin's *Cosmocorps*, 1963–64, presented a vision of future space tourism, with metallic go-go boots and gloves 15
1.3 Pierre Cardin's Winter 1966/67 collection featured helmet hats, paired here with spacesuit-white coats 17
1.4 Actress Jane Fonda wearing a form-fitted costume designed by Jacques Fonteray 18
1.5 Entrepreneur and founder of Virgin Galactic, Sir Richard Branson, poses in a Virgin-branded mock spacesuit to celebrate Virgin Money's birthday 22
1.6 Chanel stages the launch of a life-sized model rocket inside the Grand Palais museum during Paris Fashion Week, March 7, 2017 26
1.7 Kitsou Dubois, who has choreographed dance performances for weightless environments 30
2.1 Model Audrey Marnay photographed for Vogue, 1999, in a chiffon flower-print strapless dress by Emmanuel Ungaro Haute Couture 46
2.2 Viktor & Rolf Fall/Winter 2007 47
2.3 Kate Moss appears to hover weightlessly in a hologram at Alexander McQueen Autumn/Winter 2006 48
2.4 Iris van Herpen's runway show for *Biopiracy* (Fall/Winter 2014/15) 49
2.5 An underwater photoshoot by Zena Holloway for *Cent* magazine 53
2.6 Photographer Michael David Adams's underwater photographs 54
2.7 and 2.8 In Viktor and Rolf's Spring/Summer 2006 collection, Upside Down, the same gown is worn upside down (top) and the right way up (bottom) 59
2.9 A fashion model works her way down a vertical runway constructed on the side of the Marshal Fields store in Chicago, 2003 60
2.10 Peter Knapp, *Courrèges for Elle*, 1965 63
2.11 Julian Roberts's dress made using the "subtraction cutting" method 65
2.12 Julian Roberts's alternative approach to making a skirt 66

3.1 A Russian instructor (left) assists former Microsoft software developer and space tourist Charles Simonyi to don a space suit during a parabolic flight aboard a zero-gravity simulator, before his journey to the ISS in April 2007 72
3.2 Three designs for NASA's Z-2 spacesuit that were offered for public vote 74
3.3 NASA astronaut Eric Boe wears Boeing's blue spacesuit designed to be worn by astronauts flying on the CST-100 Starliner for NASA's Commercial Crew Program 77
3.4 NASA astronaut Karen Nyberg on board the ISS 85
3.5 NASA astronaut Shannon Walker, on board the ISS wearing a COTS T-shirt and cargo pants modified with Velcro strips that can be used to attach pockets and tools 88
4.1 Edward Howell's proposed design for Zero Gravity Prehensile Footwear 98
4.2 The body changes shape as a result of the effects of weightlessness 102
4.3 Japanese astronaut Koichi Wakata is pictured wearing the Penguin-3 antigravity pressure/stress suit on board the ISS 103
4.4 In microgravity, the body adopts a neutral body posture 105
4.5 A drape test measures "how fabric falls in space" 110
4.6 Diagram of an off-the-shelf polo shirt, showing primary and secondary points of direct support, at which the force of gravity causes the shirt to come into contact with the body 111
4.7 Diagram showing off-the-shelf polo short in microgravity, based on a survey of NASA videos and photographs of astronauts on board the ISS 112
4.8 Cosmonaut Valery I. Tokarev at work in the Russian-built Zarya section of the ISS 113

ACKNOWLEDGMENTS

I would like to thank Susan Buckle of the UK Space Agency for sharing her experiences of weightlessness, Nicola De Main for her advice on dressmaking terminology and her ongoing involvement in the Spacewear project, and the University of Hertfordshire for its support throughout the writing of this book and the related ongoing research.

INTRODUCTION: THE FINAL FRONTIER OF FASHION

We are entering a new, commercial space age, in which it is possible to purchase a journey into orbit or beyond, but designers are only just beginning to imagine the clothing needs of this new industry and its customers. Space presents new challenges to engineers and designers, which will need to be addressed in preparation for mass space tourism. The new space race sees small and large commercial organizations competing to provide designed objects for use by future space travelers. While private spaceflight providers such as SpaceX, Blue Origin, and Virgin Galactic develop plans to enable mass space tourism, smaller and more specialist firms, such as Final Frontier Designs, have emerged with the aim of developing individual objects that may one day be used on board those commercial spacecraft. Products ranging from cocktail glasses to sports equipment to gloves and shoes have been redesigned in anticipation of the needs of future space tourists. In this new market, there is scope for the emergence of a field of specialist practice dedicated to the design of fashion for space. While clothes of many kinds are already in use in government-funded space travel, there is currently limited activity that anticipates the fashion needs of space tourists.

In open space, the way that we clothe our bodies is essential to our survival. Space cannot be accessed "with our naked body but only through our technology: the exoskeleton or cocoon of a spacecraft, [or] a space suit" (Bureaud, 2006). Engineers have dedicated decades to developing the space suit, but far less time has been devoted to considering the features of dresses to be worn in the interiors of space vessels, where there is no need to protect the body from the dangers of open space. In the protective environment of the interior of a spacecraft, passengers are safe from the vacuum of space but are weightless. The weightless environment poses challenges to methods and processes that have long been taken for granted in fashion design and dressmaking on Earth. These methods need to be reconsidered in light of the emergence of a commercial space travel industry, which will in turn lead to the requirement to clothe future space tourists for the weightless environment.

Privately funded space travel is becoming a reality, thanks to commercial space agencies such as Virgin Galactic, Blue Origin, and SpaceX, who create commercial competition for NASA and other government agencies. NASA has contracted SpaceX and aerospace industry giant Boeing to become the first private firms to fly astronauts to the International Space Station (ISS) (Boyle, 2017), and they have pledged to make spare seats available to tourists (Dinerman, 2015). These agencies also have their goals set on entirely commercial exploration. The Commercial Spaceflight Federation (CSF) unites seventy commercial spaceflight developers, operators, and spaceports from around the globe and is working to establish a regulatory environment that will enable further growth in the sector (CSF, 2015, p. 5). NASA predicts that private space stations will be a reality by 2025, providing destinations for space tourism and private enterprise. SpaceX's successful landing of the Falcon 9 reusable rocket, dubbed the "holy grail of space flight," has laid firm foundations for their future commercial missions (Seligman, 2016). SpaceX founder Elon Musk (2016) has outlined plans for commercial interplanetary travel between Earth and Mars, following the financial model of the aviation industry, and has described his long-term goal—"to make [the journey to Mars] affordable to anyone who wants to go."[1]

The spaceflights on offer to space tourists will likely include short suborbital flights, which briefly fly high enough to offer a few minutes of weightlessness, and longer, orbital trips, which might include, for example, a stay on board the ISS (Crouch et al., 2009, p. 445; Webber, 2010, p. 1646).[2] Virgin Galactic has also proposed point-to-point flights that would transport passengers via the edge of space to one of many spaceports that are currently planned for Stockholm, the UK, and Malaysia, among others (Doule, 2014b, p. 67; UK Department of Transport, 2017; Webber, 2010, p. 1646). Elon Musk (2016) envisions point-to-point orbital travel making it possible to fly from New York to Tokyo in twenty-five minutes, launching from and landing on purpose-built offshore floating platforms. The more costly option, a long-duration stay in space, is already offered by US-based Space Adventures, whose previous clients included Dennis Tito, a private citizen who traveled into space as the world's second space tourist in 2001.[3] Space Adventures currently offers, at a cost of about $20 million, a stay on board the ISS, but plans for inflatable space hotels may provide future tourists with a more purpose-built tourist destination (Strickland, 2012, p. 925). A range of market surveys have identified enough interest in all of these forms of space tourism to suggest that they will soon be commercially viable on a large scale (Collins, Stockmans, and Maita, 1995; Barrett, 1999; Crouch, 2001; FUTRON Corporation, 2006; Crouch et al., 2009). Such surveys reveal that potential future passengers' most positive expectations of space travel include "being able to float freely in zero gravity" and "viewing space and the Earth" (Peeters, 2010, p. 1627).

Having witnessed the birth of a "nascent commercial space tourism industry" (Crouch et al., 2009, p. 442), policy makers from government and industry have begun to consider the ways in which they will need to adapt in order to accommodate the needs of the space tourist and the issues that arise as a consequence of the likely ongoing expansion of commercial space agencies. Steven Freeland (2005, pp. 10–11) outlines the ways in which legal definitions of astronauts will need to change, as Earth's obligations toward these "envoys of mankind" will need to be different from our approach to commercial "spaceflight participants." Space agencies participating in the ISS project have already ensured that their documented remit is to allow participation not only from professional astronauts and cosmonauts but also from spaceflight participants of other kinds, including engineers, educators, artists, and tourists. Those leisure travelers will need to be differentiated from professional astronauts both legally and aesthetically, in part through dress. Much like tourists on Earth, they may desire to purchase new wardrobes for their journey and showcase these new clothes in selfies or videos that are broadcast via social media to friends and family on Earth.

These new audiences are likely to pay a premium for their first space travel experience. Although Elon Musk (2017, p. 48) anticipates that space travel will become increasingly affordable thanks to reusable rockets, advance tickets for Virgin Galactic's first commercial flights have sold for $250,000 each (Wall, 2013), and SpaceX has reportedly charged in the region of $35 million to the two unnamed passengers who will be the first private citizens to travel around the moon (Crosbie, 2017). These prices locate commercial space travel firmly within the luxury market, available only to consumers with significant disposable income and, much like luxury fashion, "just beyond the realms of attainability" (Groom, 2011, p. 503). Designers who choose to cater to these audiences will need to not only consider the conditions of space but also, more specifically, how elements of luxury lifestyles could be transformed for this new environment. It is perhaps for this reason that many of the products that have already been speculatively designed for space tourists target disposable income, such as Ballantine's whisky glass for space, and the Cosmic Lifestyle Corporation's Zero Gravity Cocktail Glass.

The most significant concern for designers targeting space tourism is weightlessness. The appeal of space travel lies partly in the experience of weightlessness and the observation of "low-gravity phenomena" (Strickland, 2012, p. 293). Outside of Earth's gravity, objects behave in ways that are appealing in their strangeness, and this poses a number of new challenges for designers. Products that are taken for granted on Earth will need to be adapted or reconceived to work in the weightless environment. Designers will be required to rethink many of the assumptions about weight and balance upon which gravity-bound design is founded, leading to new approaches to product design, interiors, and fashion. Perhaps more so than any other phenomenon encountered during a journey to space, it is this weightlessness that poses the most significant challenge to fashion designers, who are accustomed to exploiting the weight of fabric to create drape, and designing for upright bodies that are rooted to solid ground.

Clothing plays an essential part in any organization's plans for space exploration. Bradley Quinn has gone so far as to suggest that clothing has not only accompanied us to space but has also enabled space exploration, observing that "the principle of a spacesuit suggested that clothing could . . . penetrate environments previously unthinkable" (Quinn, 2002, p. 3). The spacesuit enables an experience of space unlike any that would be possible on board a spacecraft. Spacesuits (specifically, Extravehicular Mobility Units or EMUs) enable astronauts to step outside of a spacecraft to partake in spacewalks and other extravehicular activities, in which astronauts are tethered to the outside of the spacecraft. Visibly awestruck by the experiences that were enabled by his spacesuit, astronaut Chris Hadfield (2014) describes the spacewalk as "the most magnificent part" of his time in space:

> You are in a one-person spaceship that is your spacesuit, and you're going through space *with* the world . . . You're not looking up at the universe, you and the Earth are going through the universe together, and you're holding on with one hand looking at the world turn beside you. It's roaring silently with color and texture as it pours by mesmerizingly next to you. And if you can tear your eyes away from that and you look under your arm down at the rest of everything, it's unfathomable blackness, with texture you feel like you could stick your hand into, and you are holding on with one hand, one link to the other seven billion people.

Hadfield is not alone in describing the spacesuit as a "one person spaceship." The requirements of the EMU are those of a "miniature spacecraft" (Jordan, Saleh, and Newman, 2006, p. 1137). It must enclose the wearer in a survivable atmosphere and protect him or her from the dangers of extreme heat and cold, radiation, and particle impacts. The image of a man in a spacesuit represents man's greatest achievements and his mastery over the conditions that have kept us earthbound. The suit is the "heroic armory which enables our encounter with the final frontier" (Shaw, 2004, p. 123).

Space, and the artificial environments that aim to replicate it, provides challenges for spacesuit engineers, and may also increasingly concern fashion designers. "A new vision is needed for the development of a wardrobe for NASA's journey to Mars in the 2030s," and prototypes of their Z-series of spacesuits offer some insight into how this new wardrobe might be designed (Orndoff, 2015a, p. 1). NASA developed the Z-2 line of spacesuits in part based on responses to a public vote that selected the final design from a number of different styles (Holpuch, 2014). The consequence of this public involvement is that NASA suits are beginning to incorporate elements of fashion. It has become necessary for the suits to mirror trends in contemporary fashion design, drawing on trends for vibrant colors and sportswear, combined with visions of the future in recent sci-fi film costumes. Commercial agencies are also vying to take the lead with clothing design to be worn in the interiors of their first manned vehicles. Virgin Galactic has employed Japanese fashion designer Yohji Yamamoto, in collaboration with Adidas, to develop flight clothing for its crew (Alvarez, 2016).

Anticipating future demand, spacesuits will soon be available off the shelf to civilian consumers, as is the case with Final Frontier Design's proposed line of ready-to-wear pressure suits (Szewczyk, 2015). The company's contract to supply counterpressure gloves to NASA illustrates a trend for the convergence of government and commercial interests in the space sector, in particular how government agencies are becoming increasingly reliant on the design skills of commercial partners. The involvement of commercial design partners, not least Final Frontier Design, has made spacesuits available for public consumption. Spacesuits are now marketed as "ultimate in bespoke couture" (Fernholz, 2015). They are no longer just uniforms for astronauts, and might one day become the ultimate wardrobe item for fashion-forward consumers.

Future spacewear designers will not, however, be required to design fully functioning spacesuits. Although the enduring image of the astronaut depicts him clothed in a spacesuit, astronauts more often find themselves in everyday clothing designed to be worn on Earth. Shielded from the hazards of open space by the shell of a spacecraft, many space travelers have no need for a spacesuit. While Space Adventures has offered a spacewalk to its ISS passengers, mass space tourism is likely to concentrate on the safer option of intravehicular experiences for all but a few of the highest-paying travelers. These travelers will spend their time engaged in onboard microgravity activities, such as "low gravity sports" (Benaroya, 2016, p. 123) or the "zero-g games," and dining that Elon Musk (2016) anticipates will occur on board his interplanetary spaceship. While on board the ISS, NASA astronauts clothe themselves in commercially available, "off-the-shelf" wardrobes of polo shirts, sweaters, and shorts or pants that were not designed with weightlessness in mind (Fratto, 2005, p. 2). It is likely that the same will be true for many space tourists, who will be able to wear a wardrobe of off-the-shelf or couture fashion, purchased for their trips much as one might for an Earth-based holiday.

While space tourists could follow the lead of NASA astronauts, wearing everyday Earth clothes for intravehicular activity, NASA and commercial space organizations recognize that more specialist spacewear is needed, for both practical and aesthetic reasons. Annalisa Dominoni (2005, p. 2) observes the consequences of wearing US-issued garments in the microgravity environment of the ISS:

> Images transmitted [from the ISS] show crew members with t-shirt collars rucked up around their necks because of lack of gravity, or with their backs

> uncovered because garments have not been designed to consider body postures, movements and physiological alterations caused by microgravity. The final effect, apart from any aesthetic considerations, is that of slovenliness not to mention the obvious discomfort of such garments.

Dominoni's observations illustrate the need for clothing to be adapted or redesigned for the weightless environment. The problems facing spacewear designers cannot be resolved by producing a simple digital or mathematical model of the way that fabric behaves in space. To develop an understanding of the effects of microgravity on drape would be a massive task, as evidenced by the ongoing experiments that NASA continues to perform with the aim of adapting other everyday products for use in space. Some headway has been made by researchers and designers with an audience of professional astronauts in mind. The VEST project sought to address these problems, as part of a "feasibility study to improve living and working conditions in space," by investigating clothing cuts and patterns suitable for the neutral posture adopted by astronauts in microgravity. It led to the development of Garments for Orbital Activities in WeightLessness (GOAL), which would adapt to the changing posture and proportions of the body in microgravity, would not fall awkwardly against the body without the force of gravity, and would permit ease of movement around the cabin of the ISS. The project focused in particular on new fibers and fabric, for use in a "second skin" that would offer astronauts more comfortable wearing experiences (Dominoni, 2015, loc. 226–240). The project recognizes the positive effects on mental well-being that are achieved when astronauts are given improved clothing choices.

Commercial passengers' demands will differ from those of professional astronauts. Future space tourists may eventually clamor for the freedom to select from a variety of styles, tailored for weightlessness just as they are for the seasons on Earth. The cabins on board commercial vessels will be spaces for social interaction, in which both verbal and aesthetic messages will be transmitted between passengers. Passengers will likely desire to express their individuality and distinguish themselves visually from crew members or professional astronauts. Thus, social interactions on board commercial spaceflights will be enhanced by the freedom to choose nonregulation clothing, designed specifically for the conditions on board the craft. Moreover, the space tourist's desires for luxury, exclusivity, and novelty, exhibited in their decision to purchase a ticket to space, may be reflected in their clothing choices. While Earth fashion has already been significantly influenced by the Space Age when 1960s fashion took aesthetic influence from the first Space Race, this existing fashion will not be sufficient to fulfill the needs of space tourists. In the twenty-first century, as commercial space travel becomes a reality, fashion will need to be designed for real-life space travel. In the New Space Age, there will be real opportunities to carry fashion into space—for fashion not only to aspire toward space travel but also to become an integral part of the space tourist's experience. Garments designed exclusively for use in microgravity will mark future spacewear apart from fashion that simply aspires toward the space aesthetic and will mark its wearers as an exclusive group whose experiences set them apart from their earthbound peers.

The effects of weightlessness necessitate a radically new approach to dressmaking and fashion design. As Selina Pang of London's Science Museum suggested in conversation with the author, designers will soon need to consider the fashion needs of future space tourists in the same way as it has previously engaged with other specialist markets, such as the activewear market. In the era of mass space tourism, a new category of fashion, *spacewear*, will need to be distinguished from Earthwear. The introduction of this new category of fashion will not only acknowledge the different conditions in which garments will be worn but also emphasize the extent to which fashion designers and apparel manufacturers will need to adjust their approach to dressmaking and design. Spacewear will need to be distinguished from all forms of Earthwear, including the aspirational fashions of the Space Age, as these historical fashions, while taking inspiration from space travel, were intended to be worn on Earth's surface.

Moreover, spacewear will need to be subdivided into categories. Just as fashion is currently divided into categories such as sportswear, daywear, nightwear, and so on, which are indicators of function and design, spacewear categories must be distinguished for different uses and environments. Engineers have already established that "there is no universal spacesuit," because suits must be designed for a diverse range of "mission needs" (Frost, 2015). The most fundamental categories will mark the distinction between spacewear for extravehicular activities (spacewalks) and intravehicular activities (on board a spacecraft). The extravehicular environment exposes bodies to such extreme conditions that the body must be wholly sealed within a spacesuit. By contrast, intravehicular spaces house artificially maintained environments, protected from the dangers of open space, where fashion designers will have more freedom to prioritize aesthetics over utility.

This book considers the developments in fashion design that must be made to meet the demands of future commercial space travelers. In particular, it considers weightless environments and how they prompt new approaches to garment design and construction. The book addresses the various features of clothes that must be reconsidered for the microgravity environments, including those encountered in space as well as on board zero-gravity flights within Earth's atmosphere. Learning from the design of spacesuits and pressure suits, as well as recent speculative fashion design, it identifies how clothes are experienced in microgravity, and how designers must accommodate these conditions. I propose that there are three things that need to be accomplished before the fashion industry is prepared to clothe future space tourists:

- Designers must appreciate the ways in which fashion design and dressmaking have, until now, been heavily informed by the assumption that garments will be worn in Earth gravity and concentrate their attentions on weight, drape, and silhouette, all of which are disrupted by weightlessness.
- Designers must identify the extent to which these assumptions need to be revisited and revised.
- Designers must identify new constraints and conditions that will inform design for weightless environments.

This book aims to support designers in these three tasks by directing future researchers and designers toward particular consequences of weightlessness that affect clothes and their wearing, and by highlighting existing fashion practices that will help designers to connect past, present, and future.

Future fashion designers will be required to reassess many of the dressmaking and design processes that are fundamental to fashion on Earth's surface. Drape, which is a staple of garment design, is defined as a product of gravity. This book asks whether drape can be said to exist without gravity and how designers' approaches to drape need to change in order to suit weightless environments. Designers must consider not only changes to the behavior of fabric but also changes in body structure. As the body adapts to reduced gravity, it adopts a neutral posture, and weight is redistributed as the upper body swells and the spine lengthens. In the long-duration space travel that is proposed for missions to Mars, these distortions will be more extreme. Garment silhouettes must necessarily compensate for the redistribution of weight around the body. The relationship between body and garment needs to be reassessed. Garments in microgravity do not hang on the body as they do on Earth. The weightless garment contains a body but is not supported by it. Garments contain the body differently in different gravitational conditions, leading to "a newly found balance between the muscles and the tension of fabrics" (Dominoni 2003, p. 278).

Notions of "top" and "bottom" will need to be discarded, as "up" and "down" hold no meaning in microgravity (D'Aloia, 2012, p. 222). While we rarely view a garment from above or below on Earth, microgravity conditions will expose these unseen surfaces, forcing designers to consider the appearance of a garment not just from the front, back, and sides, but also from above, below, and all other angles in between. The notion of the silhouette, which has been so important in defining historical developments in fashion, will also need to be redefined. For more than a century, developments in fashion have been introduced as "the new silhouette" (*Vogue*, April 1913, p. 44). Spacewear designers will need to abandon their preconceptions about the primacy of the front or side silhouette and instead begin to consider the appearance of a garment from all orientations.

Trips on board an orbiting space station, suborbital flights, and zero-gravity flights within Earth's atmosphere are already available, or planned for the near future. There is therefore an imminent need to consider the consequences of these kinds of travel for all fields of design, not least fashion. While humans may begin to build a settlement on Mars in the not-to-distant future, this book focuses on the weightless environments encountered while away from the surface of the earth or any other planet, including those that are already experienced by astronauts and space tourists. The microgravity conditions on board zero-g flights and spacefaring vessels are distinct from the low or high gravity that will be experienced at extraterrestrial colonies and bases. If we do eventually establish bases on the moon and Mars, a new set of conditions will need to be considered, beyond those that are the focus of this book.

The first chapter, "Space style," establishes the context in which the need for spacewear is emerging. It introduces the relationship between space travel and fashion, beginning with the Space Age of the 1960s and culminating in an exploration of how the contemporary art and design, including fashion, engages with weightlessness and the possibilities of commercial space travel. Engagement with the real-life concerns of space travel, such as weightlessness, sets contemporary spacewear apart from Space Age fashion of the 1960s. The chapter considers how fashion designers can learn from observations already made in other fields of creative practice, including performance art and zero-gravity product design. The chapter will identify weightlessness as a particular feature of real-life space travel that is only just beginning to be considered by designers but will become an essential concern as space tourism becomes a reality.

The second chapter, "Weightlessness on (and above) the catwalk," addresses fashion's fascination with weightlessness, as manifested on the catwalk and in fashion photography. It identifies the ways in which fashion has, often unwittingly, preempted many of the concerns of future spacewear designers. Fashion designers and retailers have experimented with suspension and orientation in ways that will help future spacewear designers to understand the effects of weightlessness of fashion. By exploring underwater photography, rigged displays, and experiments with nonstandard orientation, the chapter will show the extent to which the fashion industry has engaged (or, sometimes, failed to engage) with aspects of design and display that will be important when designing for microgravity. These practices will be presented as precursors to key principles in spacewear design.

Chapter three, "Commercial spacewear," identifies the demand for commercial spacewear that is arising as a consequence of the commercialization of space travel. Government space agencies already clothe their astronauts in commercial off-the-shelf wardrobes for everyday activity on board the ISS, and commercial space organizations have begun to develop and commission distinctive spacesuits for future space tourists, working in collaboration with fashion and costume designers. In the newly emerging commercial space age, spacesuit designers have begun to take inspiration from fashion and have begun to operate like fashion houses, producing ready-to-wear lines as well as bespoke, "couture" items. The chapter predicts the emergence of a spacewear fashion system, with concerns and structures that mirror those of existing fashion systems. It considers the constraints that must be accommodated by contemporary spacewear designers and the balance of form and function that is sought by these designers.

The final chapter, "The clothed body in microgravity," examines the microgravity conditions that exist in orbit and on zero-g flights, and considers how these environments affect the wearability of clothes. It will begin by examining how these conditions affect the relationship between the body of a space traveler and the worn garment, considering the effects of microgravity on the everyday clothing worn by astronauts. The behavior of the body, and of fabric, is observed to be distinctly different without the force of gravity acting upon it. The chapter describes how the posture that the body adopts in weightlessness, the ways in which it can be stabilized in its surroundings, and the way in which fabric encloses but does not rest upon the body are presented as key considerations for future spacewear designers.

The conclusion considers that post-gravity, post-human wearer of future spacewear and identifies the audiences that will likely serve as customers for spacewear designers and manufacturers. It summarizes the key findings of this book as a list of key considerations for future spacewear designers. This list may act as a checklist for future designers and manufacturers to ensure that their garments are suitable for microgravity environments.

1
SPACE STYLE

Half a century from the beginning of the Space Age, its cultural and aesthetic legacy persists. Fashion inspired by space travel continues to feature the metallic fabrics and form-fitting silhouettes that characterized the Space Age designs of Pierre Cardin and Andre Courrèges. Space Age fashion was as much influenced by science fiction as it was inspired by real achievements and advances, but as we enter the age of space tourism, designers must begin to consider the real requirements of spacewear, and how the weightless environment will change the ways in which we wear clothes. Artists and designers have begun to directly engage with the weightless environment, sending works to the International Space Station (ISS), or conducting experiments in the small pockets of weightlessness that occur on board parabolic flights. These experiments give us some insight into the creative explorations that may take place on board commercial spaceflights, and the new ways in which designers of the future may need to approach their practice.

The image of the hero astronaut has been superseded by the image of space entrepreneurs Elon Musk and Richard Branson, whose privately funded ventures will determine the direction of the commercial space race. These new space heroes are driven by the same pioneering spirit as Kennedy and his elites at NASA, but armed with the freedom that comes with self-finance. The new commercial space age offers opportunities for creatives, from performance artists to product designers, to adapt their practice for the weightless environment. Catering to future space tourists and other explorers of the new frontier, these practitioners recognize "post-gravity" experience as a gateway to new approaches to designed artefacts and related cultural practices. A field of design is emerging to cater to the everyday experiences that are defamiliarized by weightlessness. Within this broader field of design for weightlessness, there is scope for designers to revisit Space Age fashion and redefine it for the final frontier.

The first Space Race

In 1957, the launch of the Sputnik satellite by the Soviet Union marked the dawn of the Space Age. The Soviet Union excelled in space, beating the United States to launch the first man, Yuri Gagarin, into space in 1961, and the first woman, Valentina Tereshkova, in 1963 (Burleson, 2005, p. 224). Concerned about the political and military ramifications of falling behind the Soviets at the height of the Cold War, the United States embarked on their own program of space exploration, prompting the start of the first Space Race. In 1958, James R. Killian, science advisor to President Eisenhower, presented a report outlining four arguments for pursuing an American space program. First, it described Man's "compelling urge to explore and discover," the common value that would unite American citizens in support of the program. Second, it identified a "defense objective," fueled by concerns about Russia's military capabilities. Third, it described the "national prestige" that could be gained by overtaking Russia's lead, and fourth, the opportunities to advance scientific knowledge (Parker, 2009, p. 93). The United States began to balance the scales, but constantly trailed behind the Soviet Union by about a year, launching their first satellite, *Explorer*, in 1958, and transporting John Glenn into space in 1962 on board the *Friendship 7* rocket.

Running as a presidential candidate, John F. Kennedy's criticisms of Eisenhower frequently included reference to the inadequacies of the existing space program (Jordan, 2003, p. 211). Kennedy and his advisors conceived a new space program primarily as a show of military strength. Communications from Kennedy referred directly to competition with the Soviet Union. In a memo to Vice President Johnson in April 1961, he expressed anxieties over the Soviet Union's superior space program, and an eagerness to invest "maximum effort" into "beating the Soviets" to the Moon. Kennedy's space program was all about "winning"; according to Jerome Weisner, chairman of Kennedy's Science Advisory Committee, the "rockets were a surrogate for military power" (Logsdon, 2010, p. 83).

Recognizing the importance of public support of his ambitions for a Moon landing, in 1962 Kennedy gave an address at Rice University that presented the lunar program as both plausible and urgent (Jordan, 2003, p. 210). The speech established the Moon as an achievable, tangible destination, crafting a "spatial rhetoric that minimized the distance and obstacles" (p. 215). This required the presentation of the Moon, not as the ultimate destination, but as the first step toward much more significant exploration of deep space. Referring to the "pioneering heritage" of the United States, Kennedy presented America's space program as a manifestation of the nation's shared values (p. 213). Space was, he declared, a "new sea" upon which "we set sail," all the while stressing that this was to be a peaceful journey that could be taken "without feeding the fires of war." Space should not be, he asserted, "filled with weapons of mass destruction, but with instruments of knowledge and understanding" (Kennedy, 1962).

Early 1960s American national mythology was shaped equally by notions of the old and new frontier. Kennedy's New Frontier speech prompted comparisons between the Wild West and the Space Race. Popular film resurrected frontier adventures, which played in cinemas alongside science fiction. The "Space Western" emerged as a genre that united these two frontiers, relocating the mythologies of America's past into visions of its future (Montheith, 2008, p. 20). For Kennedy, the "final frontier" was more than space. The metaphor was intended to represent the changes and challenges that would set the 1960s apart from the first half of the twentieth century. The 1960s would be a decade of "uncharted territory" in many ways: in Kennedy's own words, "A frontier of unknown opportunities and perils," including many of terrestrial origin. The "New Frontier," he declared at the 1960 Democratic National Convention, was "a set of challenges" that would face the president and his American people (Logsdon, 2010, p. 8). The "conquest of space" seemed to provide evidence that even the greatest challenges could be overcome (Kohonen, 2009, p.122). It represented "a victory over time and place as transcendence of all earthly hardship" (p. 116).

The Apollo astronaut was a new breed of hero for the Space Age, presented by Kennedy as "a unique blend of frontier adventurer and scientist" (Jordan, 2003, p. 221). Astronauts (and cosmonauts) "had the role of new, quintessentially utopian, heroes" and this status was expressed in the "heroic armoury" of the spacesuit (Kohonen, 2009, p. 123; Shaw, 2004, p. 123). Clothes played a significant role in the lunar missions, both practically and symbolically, enabling Man to set foot on the Moon, while symbolizing the technological advancement of mankind.[1] Clothed in a spacesuit, Man had evolved into spaceman. One image "captured . . . the euphoric liberation of a species coming of age," and cemented the image of the spacesuit-clad hero in the popular imagination (Wachhorst, 2000, p. 40). In NASA photographs from 1965, astronaut Ed White is seen taking Man's first spacewalk, drifting above the curved blue horizon of the Earth, tethered by a gold umbilical line to the Gemini 4, enveloped from head to toe in white, with the Pacific Ocean reflected in his gold-plated visor (see Figure 1.1). Clothed his spacesuit, he is more than human, liberated from the gravity that binds man to the Earth's surface, and able to witness all of mankind from his orbital vantage point.

While the popular imagination was captivated by such images, designers contributed to the Space Age aesthetic with streamlined forms and cosmic motifs. Yonel Lebovici's *Satellite* table (1965) and *Soucoupe* (saucer) floor lamp (1970) featured glowing acrylic discs inspired by the cosmos (Fiell and Fiell, 2005, p. 415); Googie architecture delivered flying-saucer shaped buildings including Edward E. Carlson's Seattle *Space Needle* (1961) and Pereira & Luckman's *Theme Building* at the Los Angeles International Airport (1961). The fashion industry sought modernity, and did so through references to youth, activewear, and the Space Race (Guillaume, 1998, p. 17; Baldaia, 2015). "White was everywhere," in interiors, product design, and fashion (Guillaume, 1998, p. 13). For spacesuit engineers, white was the color of choice to ensure that astronauts could remain clearly visible against the black backdrop of space,

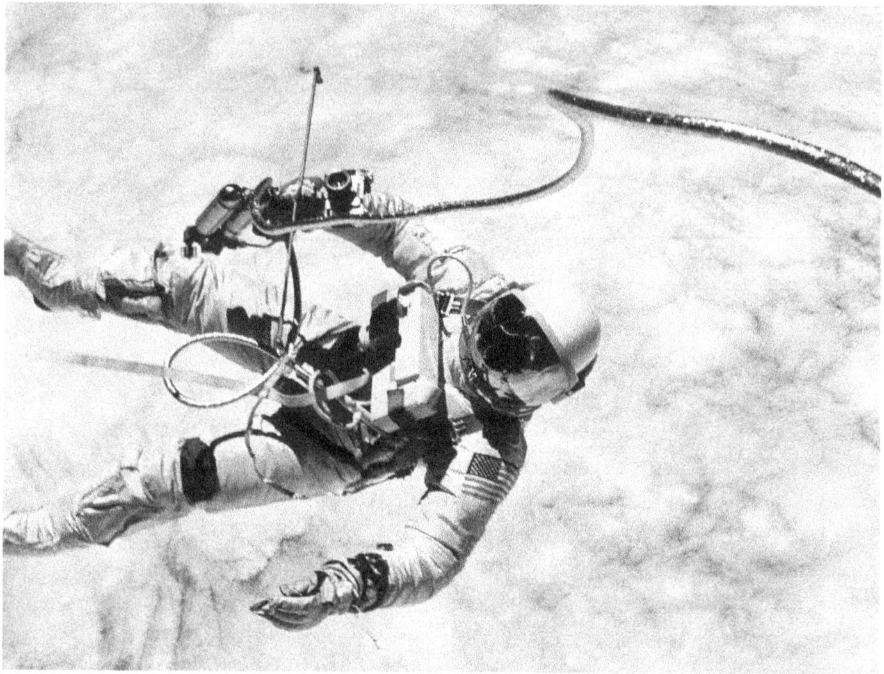

Figure 1.1 Astronaut Edward H. White II floats above the Earth on the first spacewalk. White is attached to the spacecraft by an umbilical line and tether. The visor of his helmet is gold plated to protect him from the unfiltered rays of the sun. © Jim McDivitt/NASA/ Roger Ressmeyer/Corbis/VCG

but for designers like Andre Courrèges white not only signified space travel, it also enabled them to emphasize structure over surface (Pavitt, 2008, p. 54). In fashion, architectural forms emphasized a "new functionality," with "lean, angular silhouettes" made possible by new materials and new construction techniques inspired by engineering (p. 50).

Jane Pavvitt (2008, p. 9) notes the irony of this optimistic, future-facing design, which emerged despite Cold War anxiety. With the world on the brink of destruction, designers and consumers "enthusiastically embraced" design that seemed to foreshadow a utopian future (Baldaia, 2015). For the fashion industry, and designers more broadly, "space travel provided a highly evocative source of imagery" (Sparke, 1993, p. 100). In spaceflight themed fashion spreads, models stood proud, with eyes cast upward in an aspirational gaze and, at other times, depicted as if floating through space. Suzanne Baldaia (2015) cites such imagery as contributing to a code that was established in fashion design and in the magazines that revealed Space Age designs to the public. The language of accompanying articles reinforced Space Age connotations with space travel terms and metaphors, contributing to a visual and verbal code of Space Age signifiers that linked modernity and space travel. Space Age fashion, along with the commentary that surrounded it, seemed to embody the modernist narrative of human progress that Alexander Geppert (2012, p. 9) locates within a wider "astroculture"—images and other cultural artefacts that "ascribe meaning to outer space while stirring the collective imagination."

Although the Soviet Union and United States led the Space Race, Connikie (2007, p. 46) observe that it was European designers that were most significant in the development of Space Age fashion. Pierre Cardin, André Courrèges, and Paco Rabanne established the Space Age fashion aesthetic, with spacesuit-whites and metallic, helmet-shaped hats, jumpsuits, minidresses and flat footwear. Pierre Cardin's *Cosmocorps* line (1963–64, see Figure 1.2) seemed to reflect the possibility of a future space tourism industry. Meanwhile, Courrèges's collections of the mid- to late-1960s "embody the myth of the future and the conquest of space" (Dabramo, 2012, p. 2). Courrèges believed that fashion

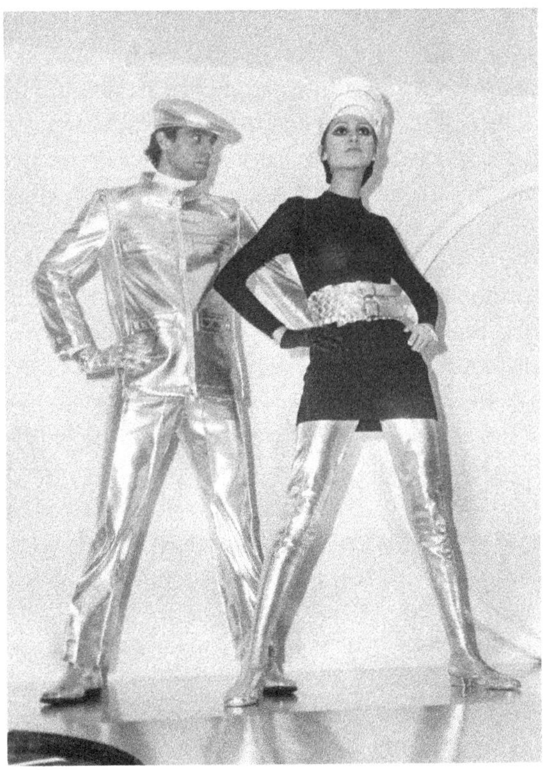

Figure 1.2 Pierre Cardin's *Cosmocorps*, 1963–64, presented a vision of future space tourism, with metallic go-go boots and gloves. Here, the models are posed with feet apart and aspirational gazes, mirroring the poses adopted by NASA astronauts. © Keystone-France/Gamma-Keystone

could be "revolutionized through new materials" (Guillaume, 1998, p. 15). Sourcing specialist fabrics from aviation and sports sectors, he began to incorporate cellulose acetate into his designs (Horwell, 2016). Newly available optical brightening agents enabled his use of brilliant whites, which appeared fluorescent in the daylight, inspired by "the bright side of the Moon" (Dabramo, 2012, p. 2).

Courrèges pioneered the A-line minidress, and in doing so separated the body from the garment, making the garment feel weightless. Feeling that an emphasized waistline cuts the body in half, Courrèges sought to unify the top and bottom of the body into a single whole. His minidresses fell from the shoulder, flaring outward toward the hem. The effect, as he saw it, was of clothes that "float," as if weightless, separated from the body so that the wearer cannot feel the sensation of wearing (p. 7). In a photoshoot for *Elle* in 1965, Peter Knapp photographed Courrèges's *Moon Girl* collection on models that appear to float weightlessly across the page (see Chapter 2).

Space Age fashion reflected a "desire to match ourselves to the images that surround us" (Shaw, 2008, p. 21). Driven by a desire to replicate astronaut style, fashion borrowed heavily from the images that NASA distributed to the press. Dressed in spacesuit white, models were photographed with the skyward gaze of Apollo astronauts. Duplicating the images of the first Moon landing, they posed with their feet apart, planted firmly on the floor in white or silver go-go boots. Faces peered out from inside helmet hats, which took a variety of forms from Emilio Pucci's 1965 plastic bubble to Pierre Cardin's helmet hats of Fall/Winter 1966/67 (see Figure 1.3). These images were aspirational, but also illustrate the extreme difference between the flights of fancy of Space Age fashion and the bulky, uncomfortable, and unflattering spacesuits worn by real astronauts. Helmet hats embodied Space Age fashion's unfulfilled aspirations, signifying space travel while serving no functional purpose. While such designs allowed public citizens of the Space Age to engage with the Space Age aesthetic, they also marked a distance between the exclusive world of the professional astronaut and the earthbound lifestyle of the civilian population.

Since there was no requirement for these garments to function as spacewear, Space Age fashion could borrow equally from fact and fiction. There needed to be no distinction between styles borrowed from NASA and those borrowed from the science-fiction narratives that also increased in popularity at this time. Indeed, Arthur Woods (Pocock, 2012, p. 336) goes so far as to suggest that it was via science fiction that the general public came to understand space exploration. Science fiction had been responsible for introducing the citizens of the world to the possibility of spaceflight (Wachhorst, 2000, p. 46). When Man set foot on the Moon, reality and science fiction converged. Since the 1930s, "the present had been a permanent progression towards the future . . . but now the future was something that was comprehensible to all" (Kohonen, 2009,

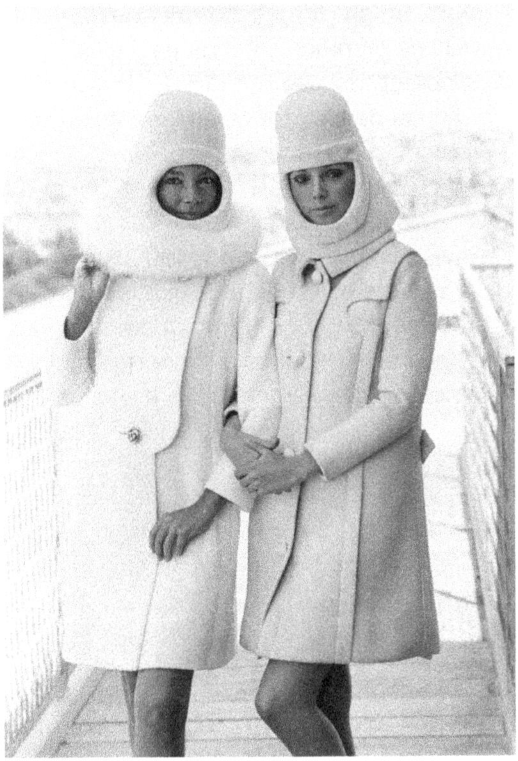

Figure 1.3 Pierre Cardin's Winter 1966/67 collection featured helmet hats, paired here with spacesuit-white coats. © Francois Pages/Paris Match

p. 123). Feats that had previously seemed impossible were now proven feasible while, at the same time, science fiction referenced real discoveries and events. Spectacular fictional visions of the future seemed increasingly plausible. In 1976, NASA indulged science-fiction fans by playing the *Star Trek* theme tune as they unveiled their first shuttle, *Enterprise*, named in homage to the show's fictional starship (Woods, 2009, p. 31). The blurring of science fiction and prediction was such that, in the popular imagination, no distinction was made between the reality and fiction of space travel. We had entered what Constance Penley (1997) refers to as the "NASA/Trek" era, in which it was feasible to imagine a path that would lead the human race directly from NASA's today to the tomorrow that is depicted in *Star Trek*. In fashion too, NASA and *Star Trek* had a close relationship. The Starfleet uniforms worn by the fictional crew of starship *Enterprise* were "based on sketches and ideas supplied by NASA scientists" (Connikie, 2007, p. 44).

The extent to which Space Age fashion was inspired by a convergence of reality and science fiction is perhaps never more apparent than in Jacques Fonteray and Paco Rabanne's costumes for Roger Vadim's *Barbarella* (1967).

Fonteray and Rabanne's designs for the movie's spacefaring heroine, played by Jane Fonda, bore much in common with the rigid plastic and metal garments of Rabanne's fashion collections (see Figure 1.4; Lundén, 2016, p. 186). In the film's title sequence, Fonda performs a weightless striptease. As the sequence begins, she is presented as an anonymous silver-suited astronaut, clothed in a bulky costume that gives her the silhouette of any of NASA's astronauts. As she sheds the suit, piece by piece, she reveals the smooth contours of a female body, and as the reflective layer of her visor lowers, at a tantalizingly slow pace, her sultry gaze peers out. Though the film is celebrated for Barbarella's extravagant wardrobe, this is only one of two spacesuits that she wears in the film, recognizing that the future of space couture is not in spacesuits but other spacewear that will be worn where spacesuits are unnecessary.

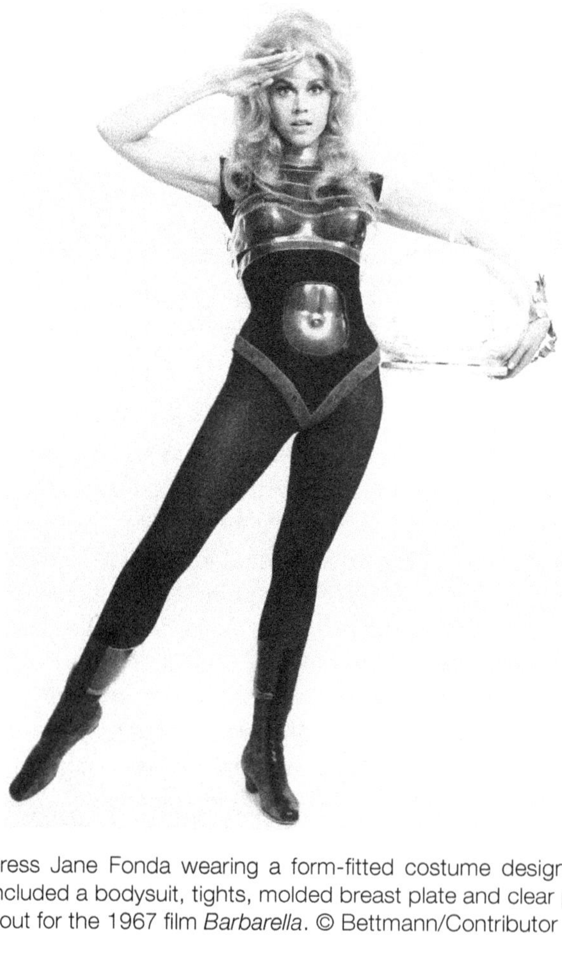

Figure 1.4 Actress Jane Fonda wearing a form-fitted costume designed by Jacques Fonteray. This included a bodysuit, tights, molded breast plate and clear plastic helmet in a publicity handout for the 1967 film *Barbarella*. © Bettmann/Contributor

Not all Space Age fashion was superficial in its association with space travel. The Space Race also had some very practical influences on fashion, and vice versa. Before the Space Race, the fashion industry had already begun to establish a relationship with aviation and, later, spaceflight. Manufacturers had identified convergences in the demands of corsetry and suits for high-altitude flight. The application of pressure on the body's exterior was key to both industries. Aviation "garments using mechanical pressure" were employed "to protect elite pilots from acceleration and depressurization in flight" while, at the same time, innovations in corsetry led to structured foundation garments that incorporated many of the same materials. "Not coincidentally, the same companies produced both garments—companies with experience manufacturing corsetry found themselves well-equipped to produce pressurized clothing for others kinds of consumers" (Hersch, 2009b, p. 349). Companies such as the Spencer Corset Company, Playtex, and the David Clark Company were involved in the manufacture of both women's foundation garments for retail and aviation pressure suits for clients including NASA (p. 346).

Nicholas de Monchaux recounts how essential the undergarment industry was to the development of the spacesuit—in particular, seamstresses accustomed to working with materials and requirements that were similar to those of NASA engineers. While NASA struggled, and failed, to develop a better spacesuit, Playtex seamstresses were able to employ the "couture handicraft" that was missing from NASA engineers' skillset. Ultimately, argues de Monchaux, NASA's early success in spacesuit manufacture was the result of "couture sewing, . . . stitching and biasing and all of the very different and special modes of expertise that fashion has always had in designing for the body" (Van Dusen, 2012). The undergarment industry seemed poised to stand alongside the aviation industry at the forefront of space travel technology as the first Space Age began. However, the ambitions of NASA's commercial suppliers were not yet enough to ensure the continued expansion of the space sector when governmental organizations began to lose faith and focus.

The commercial space age

As with all fashions, the Space Age aesthetic gradually declined; the optimism and enthusiasm of the Space Age gave way to "an era of space fatigue," and the 1970s saw a more down-to-earth approach to fashion and design (Geppert, 2018). At the same time, the Space Race lost momentum, as the implausibility of Moon colonies and long-distance space exploration became disappointingly evident.[2] As in previous endeavors, the Soviet Union was several steps ahead of the United States in the development of their first space station. The Soviet Union launched the *Salyut* in 1971, but a string of bad luck followed, beginning with a failure in the docking system that meant a delay in habitation and, later, the death of its crew as their reentry capsule depressurized on their return to Earth (Ivanovich, 2008, pp. 104; 306). Failures plagued American space shuttles too, some of them fatal. NASA remained cautiously optimistic about the future expansion of the program until the Challenger disaster of 1986, in which an explosion destroyed the shuttle, along with its seven crew, seventy-three seconds into the flight (Woods, 2009, pp. 36–40). The United States had more success than the Soviet Union with their Skylab station, which remained in orbit from 1973 to 1979, but also initially encountered problems. The station was damaged on launch, requiring a replacement heat shield before it could be inhabited (Hitt, Garriott, and Kerwin, 2008, p. 138). The Soviet Mir station, assembled in orbit over a ten-year period starting in 1986, and deorbited in 2001, was constructed of six secondary modules surrounding one core module, showing the potential for infinite expansion of future stations. It too suffered some accidents, including a fire caused by a faulty oxygen generator and a hull breach resulting from collision with a service module, both in 1997 (Oberg, 1998, pp. 30–31). Despite these setbacks, the crew achieved all of their stated goals, including spacewalks and experiments. The many successes and failures of previous stations informed technology and processes on board the ISS (Iannotta, 1997, p. 35).

The iconic image of the suited astronaut, seemingly able to withstand anything that the universe could throw at him, lost some of its potency when, in 2003, six astronauts were killed as the shuttle Columbia overheated on reentry (Tate, 2013). The human cost of the Columbia disaster was reported on the cover of several newspapers, accompanied by photograph of an empty, scorched helmet, with a missing visor. The image of the helmet was a brutal reminder of "the vulnerability of the human body" (Shaw, 2004, p. 129), and of the fact that even the greatest minds in the world cannot develop a suit that protects the human body from all of the many unpredictable threats posed by space travel. In many more positive settings, such as science museums and displays celebrating Man's conquering of space, the empty suit becomes a proxy for the absent astronaut. Lunar landing exhibitions such as the display at the NASA Space Center in Houston, Texas, use empty spacesuits as stand-ins for astronauts. Their reflective visors enable viewers to peer into an empty helmet and see their own faces gazing back, as if projected into the empty suit. In images following the Columbia disaster, the empty spacesuit has much more tragic connotations. The scorched helmet, and the detached sole of one astronaut's boot, found strewn among the debris in a field in Texas, illustrate the fate of the Columbia astronauts, the shattered visor and scattered pieces becoming a metaphor for the broken body.[3]

As the space program declined, it became clear that government space agencies were not fully equipped to fulfill their ambitions. Alongside the gradual realization that space travel would develop more slowly than the optimists of the Space Age had predicted, came the acknowledgment that no single government agency had the resources to achieve their goals alone. It would take input from commercial organizations to reinvigorate the space sector. The commercial space race would provide, among other things, new markets for spacesuit engineers, and new inspiration for fashion designers. The potential for collaboration with commercial organizations had first to be demonstrated through collaboration between national space agencies. As NASA's budgets became increasingly unsustainable, and relations between Russia and the United States improved toward the end of the Cold War, the United States and Soviet Union combined resources for their first collaboration on the Apollo-Soyuz Test Project, during which a NASA Apollo Command/Service Module docked with the Soviet *Soyuz 19* spacecraft in 1975. The project was symbolically significant, marking a transition from competition to collaboration. The launch of the Mir space station, followed by the launch of the first stage of the ISS in 1998, signaled a new era of international collaboration (Brennan and Vecchi, 2011, p. 33). At the time of writing, the bond between national space agencies is so strong that NASA is able to operate without its own shuttles, relying on the Russian *Soyuz* to ferry NASA crew to the ISS (Doule, 2014b, p. 66). Reliance on Russian transport is one of the many reasons NASA has turned to commercial partners (Bolden, 2015).

One inevitable consequence of commercial partnerships was space tourism. As early as 1979, NASA anticipated eventually running shuttles "along similar lines to a commercial airline" (Woods, 2009, p. 38). Pan Am Airlines had announced plans for commercial flights to the Moon in 1968, but it soon became evident that these plans were overambitious. By the 1970s, interested parties had revised their ambitions, instead setting their sights on the more realistic goal of orbital space tourism (Chang and Chern, 2016, p. 534). It was not until 2001 that Dennis Tito became the first space tourist. The lack of spacefaring technology in the private sector meant that Tito's trip had to be led by a national agency, the Russian Federal State Agency (Roscosmos). Roscosmos' decision to accept a tourist on board the ISS was financially motivated. Significant contributions to its space program have come from the sale of space on *Soyuz* flights to the ISS, primarily to NASA, but also to seven tourists between 2001 and 2009 (p. 538). These first space tourists included Guy Laliberté, cofounder of Cirque du Soleil, and Charles Simonyi, lead developer of Microsoft Office, who has visited the ISS on two separate occasions.

The final decades of the twentieth century saw a gradual shift in the perception of the purpose of space travel. The ideal of space travel for exploration and scientific advancement gave way to the realization that space could be exploited for commercial gain (Woods, 2009, pp. 40–41). Private enterprise now saw what Peter Dickens (2009, p. 66) describes as "capitalism's potentially infinite expansion into the cosmos." Current commercial space activities focus on launch and supply. Both national and commercial organizations have the capability to manufacture satellites, but fewer have the facilities to launch them, and so are reliant on "launch services" (Brennan and Vecchi, 2011, p. 41). SpaceX is among a handful of companies who are able to launch satellites into space, and their Dragon cargo ship has been delivering supplies to the ISS since 2012. Boeing, SpaceX, and Orbital ATK have all developed close ties with NASA.

Many of these new commercial ventures are led by entrepreneurs, whose pioneering spirit echoes the sentiment that fueled NASA's lunar program (Brennan and Vecchi, 2011, p. 20). At the time of writing, there are three entrepreneurs whose names dominate discussions about the future of orbital and suborbital space tourism: Richard Branson, Elon Musk, Jeffrey Bezos, and Robert Biggelow. Richard Branson's Virgin Galactic has positioned itself as a "global leader in suborbital tourism" (see Figure 1.5). Although their first *SpaceShipTwo* crashed in the Mojave desert in 2014, killing its test pilot, the second, the *VSS Unity*, completed successful test flights in 2016 (Berger, 2016).

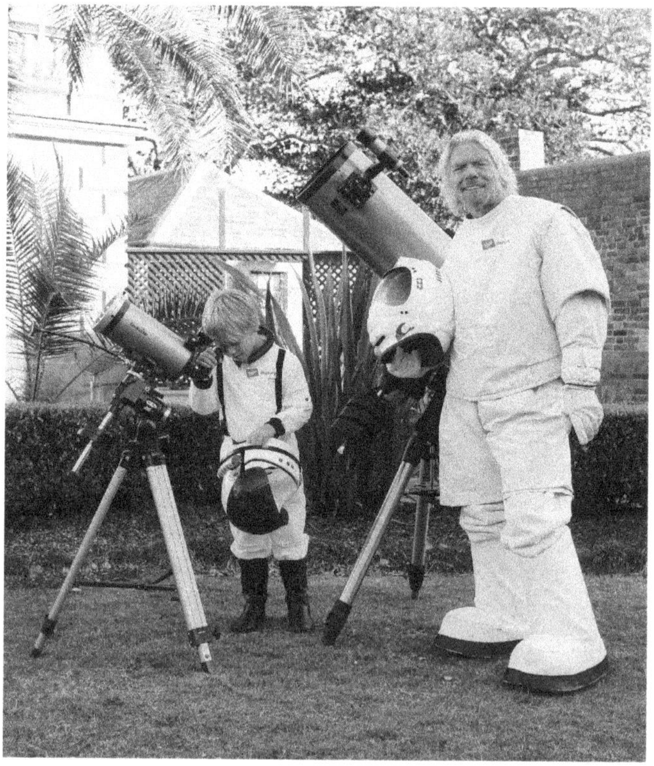

Figure 1.5 Entrepreneur and founder of Virgin Galactic, Sir Richard Branson, poses in a Virgin-branded mock spacesuit to celebrate Virgin Money's birthday. Branson announced a competition that would offer the winner a trip into space on the Virgin Galactic. © Don Arnold/WireImage

Elon Musk's SpaceX has established a strong relationship with NASA that will ensure its continued survival for the foreseeable future (Fecht, 2016). Although they too have some history of failed launches and landings, they have overcome setbacks to become the first private company to successfully return a spacecraft to Earth from low-Earth orbit.[4] In December 2017, Amazon CEO and founder of Blue Origin, Jeffrey Bezos, successfully flew a mannequin to space and back on its reusable *New Shepard* capsule, in preparation for transporting space tourists in the same vehicle (Davenport, 2017). While Virgin Galactic, Blue Origin, and SpaceX have focused on transportation, space tourists' accommodation may eventually be provided by Robert Bigelow's Bigelow Aerospace, which has already tested its inflatable "space hotel" (Bigelow Expandable Activity Module) as an additional module for the ISS (Aron, 2016).[5] These organizations are competing with established aeronautics industry giants in what has become a new, commercial space race.[6]

The private sector has reinvigorated plans for further exploration and eventual colonization of other planets, likely involving further collaboration between commercial and government organizations (Brennan and Vecchi, 2011, p. 53). Increased awareness of the damage that mankind has done to Earth, and of the vulnerability of a species that is reliant on the resources of a single planet, has led to the assertion that further space exploration, leading to the establishment of colonies on other planets, will be essential for the preservation of human civilization (pp. 30–31). The attention of the space exploration community, both national and commercial, has been focused on Mars as a potential destination, and NASA have worked with Airbus to develop the Orion spacecraft, designed to transport humans onto the surfaces of asteroids and, eventually, to Mars. In 2015, NASA outlined plans for a journey to Mars that are heavily dependent on their relationship with commercial organizations. They describe a "collaborative effort" to move from "Earth reliant" exploration, mostly in near-Earth orbit, to "Earth independent" activities that will involve harvesting Martian resources to support missions on Mars and its moons (NASA, 2015b, p. 7). While Mars orbiters and rovers have mapped the surface of the planet (p. 9), commercial partnerships will be vital in advancing to manned missions. NASA already "relies on commercial services" for "economically efficient exploration," and plans to "leverage . . . U.S. industry" (p. 15).

The most feasible prospect for a manned Martian landing comes from SpaceX (Musk, 2016). Elon Musk has acknowledged that the founding of SpaceX was largely due to his desire to contribute to the colonization of Mars (Howell, 2017a). As Musk delivers success after success in his numerous ventures, not least SpaceX, his dreams seem increasingly achievable. In September 2016 Musk outlined his plans for the colonization of Mars, making use of a reusable carbon-fiber interplanetary vehicle that will significantly reduce the cost of the journey. NASA has voiced its support for SpaceX's plans, acknowledging that its Red

Dragon capsule is likely to reach Mars a decade before NASA's own craft (Clark, 2016). NASA and SpaceX may have additional competition from other private and national organizations. The China National Space Administration, whose military status prevents it from collaborating with NASA, has also announced that they are embarking on their own Mars exploration program, beginning with the deployment of a Mars rover (Russon, 2016).

If the race to Mars proceeds as planned, it will represent the next giant leap for mankind, and will mark a continuation of the pioneering attitude that fueled the Apollo program, combined with an entrepreneurial spirit of commercial enterprise. Future manned space exploration will continue the work of the lunar missions, further extending human civilization to extraterrestrial environments. NASA (2015b, p. 1) has declared that its purpose is "like the Apollo program, [to] embark on [a] journey for all humanity." Evoking the rhetoric of Kennedy's New Frontier, their proposal refers to the mission as a "pioneering endeavor." Their plans for Mars involve not only landing, but also colonization, and they are already "developing the capabilities necessary to get there, land there, and live there"; to "work, learn, operate and sustainably live beyond Earth for extended periods . . . leading to a sustainable presence in deep space" (pp. 1; 3). These plans, if implemented successfully, will lead to the establishment of a Martian colony, a new Martian civilization that will necessarily function differently to their ancestral home on Earth, and will make humans the first "multiplanetary species" (Musk, 2016).

As companies such as SpaceX prove the viability of space travel for profit, other private and commercial ventures also seem increasingly viable. The activities of large commercial space organizations are part of a larger democratization of space exploration, and widespread access to spacefaring technologies by civilians. The last few decades have seen "an increasingly participatory ethos" in the space industry, both in the activities and policies of governmental organizations, and within the civilian sector (Kaminski, 2016, p. 221). While large corporations plan large-scale space exploration programs, opportunities are beginning to arise for independent designers and engineers to support and direct future space operations. Civilians can participate in data recording and analysis, as well as software and hardware development, in support of larger organizations, including NASA (Barschke, Ozkan, and Johnson, 2012, p. 2). NASA's own Open Government Directive uses open source software and shares data with the public via a cloud computing platform. Through small trials with public participation in 2000, NASA learned the value of volunteer labor in tasks that did not require specialist expertise. They have rolled out a larger-scale participation initiative, inviting members of the public to make contributions to NASA research via the Participatory Exploration Office (NASA, 2010, pp. 4; 92). Thanks to such initiatives, the previously closed world of astronauts and astronomers now seems accessible.

Brennan and Vecchi (2011, p. 32) anticipate that new "supermarkets" will transform space travel into a profit-seeking industry. They identify four key opportunities for private and commercial activities in space: satellite launch and manufacture for GPS navigation systems, broadcast and broadband services, and Earth observation; mining of resources such as metals, fuels, and solar energy; manufacturing opportunities; and tourism. New and expanding corporations will be able to take advantage of proposed commercial spaceports in Curacao, Sweden, Malaysia, and other locations around the globe (Doule, 2014b, p. 67). These activities will pave the way for businesses unrelated to aerospace to profit from space travel. Andrew M. Thorpe (2009, p. 27) lists potential commercial space station clients, including not only biotech companies hoping to develop pharmaceuticals and researchers conducting pure science experiments, but also creative organizations such as television and film production companies, with the aim of producing music videos, educational and reality shows, and science-fiction movies. The potential for the creative industries' involvement in space activities presents orbiting space stations as interdisciplinary venues, where innovation can be technological, cultural, or aesthetic.

Recognizing the renewed interest in space travel and the increasing likelihood of further manned space exploration, fashion designers have once again begun to incorporate space travel signifiers into their collections, looking beyond Earth for inspiration and, in particular, at manned interplanetary travel. At Paris Fashion Week 2017, Chanel staged the launch of a life-size model rocket inside the dome of the Grand Palais museum (Weitering, 2017), launching their Fall/Winter collection that features astronaut prints, large circular collars that resemble the locking ring that connects a spacesuit to its helmet, and vinyl textures that imitate the surfaces of unexplored planets (see Figure 1.6). In many ways the collection draws inspiration from Space Age fashion, with silver go-go boots and black-and-white minidresses (Yotka, 2017). Like fashion of the Space Age, the collection celebrates space travel imagery without consideration of its practical realities.

Though Chanel does not, with these space travel references, hint at any future involvement in spacewear design, the show's imagery does illustrate the revival of an interest in space travel within the fashion community. Lesser-known fashion brands are also beginning to exploit connections with the new space age, not in efforts to clothe space tourists but, much as they did during the first Space Age, to aesthetically align themselves with activities at the forefront of human innovation. Accessories label Sprayground, for example, has partnered with Buzz Aldrin on a "Mission to Mars" line that includes a white-and-grey hooded parka with panels resembling those of a futuristic spacesuit (Chua, 2017). Such examples appear to provide evidence that, with this commercialization of space travel, fashion designers will find motivation to once again look beyond Earth for inspiration. The next logical step will be to consider the realities of fashion for

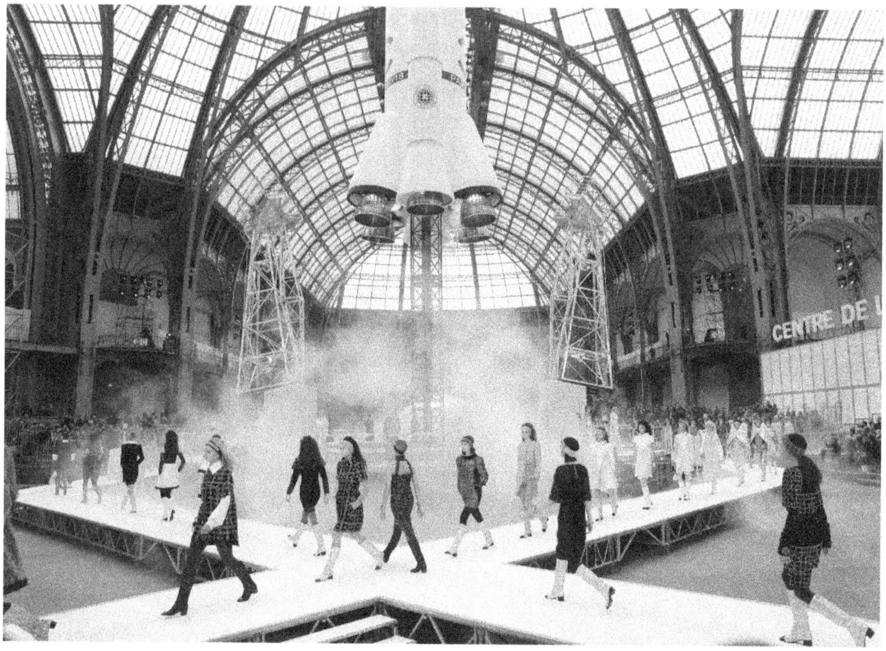

Figure 1.6 Chanel stages the launch of a life-sized model rocket inside the Grand Palais museum during Paris Fashion Week, March 7, 2017. The collection featured references to manned space exploration, including astronaut prints, and revived features of Space Age fashion including white and silver go-go boots. © PATRICK KOVARIK/AFP/Getty Images

space, and "the possibilities of space couture, as opposed to clothes influenced by space" (Timmins, 2010, p. 186). There is now a very real possibility that future consumers will take trips into orbit, or to the surface of other planets, and will need to dress for the occasion.

The art and design legacy of the Space Age

The legacy of the Space Age has been technological, political, and aesthetic. One of the most profound legacies is the altered sense of perspective that has resulted from photographs of the Earth from afar. During the Apollo 8 mission to the Moon in 1968, astronaut William Anders captured a photograph of the distant Earth rising above the horizon of the Moon that has come to be known as *Earthrise*. This image of "Spaceship Earth," so clearly a fragile store of finite resources, prompted the establishment of the first global environmentalist movement (Deese, 2009, pp. 70–71). Several decades later, as it left our solar system in 1990, Voyager 1 transmitted a "family portrait" of Neptune, Uranus,

Saturn, Jupiter, Earth, and Venus. In the image, the Earth appears as a "pale blue dot": a miniscule spec in a "vast cosmic arena." "There is," writes Carl Sagan (1994, p. 7), who was instrumental in the taking of the "blue dot" photograph, "no better demonstration of the folly of human conceits than this distant image of our tiny world":

> Our posturings, our imagined self-importance, the delusion that we have some privileged position in the Universe, are challenged by this point of pale light. Our planet is a lonely speck in the great enveloping cosmic dark. In our obscurity, in all this vastness, there is no hint that help will come from elsewhere to save us from ourselves . . . it underscores our responsibility to deal more kindly with one another, and to preserve and cherish the pale blue dot, the only home we've ever known.

The image of the small "blue dot" of Earth, as a container for all of human activity and culture, and dwarfed by other planets, exposes the relative insignificance of human intellectual and creative endeavors. Space arts initiatives have arisen from a desire to extend the reach of human culture into space. In 1969, Apollo 12 carried *The Moon Museum*, a small ceramic wafer displaying miniature works by prominent artists, to the lunar surface.[7] The Voyager probe, launched in 1977, carries with it a Golden Record containing images and music designed to communicate aspects of human culture to any spacefaring extraterrestrials that might encounter the probe on its interstellar journey. Such projects illustrate a human drive to disseminate human culture beyond Earth's atmosphere. This has evolved into a drive not just to transport records of Earth-made art into space, but also to create art that "engages outer space materially" (Kac, 2005, p. 22), that is, art that is intended for and directly affected by the conditions of space. In 1993, author and artist Arthur Woods produced an angular kinetic sculpture entitled *The Cosmic Dancer*, which was set afloat by cosmonauts on board the Mir Space Station in an effort to "promote the cultural use of near Earth orbit" (Pocock, 2012, p. 336). Woods (1993, p. 297) predicts that "as human civilization is extended beyond the planet, its culture will be altered by the conditions of the new environment," including our approach to the design of three-dimensional forms. His sculpture was first displayed throughout the 1980s on Earth at a range of different orientations, resting on different vertices but, when released into the space station, was freed from the need for a "resting point" (p. 299).

The Cosmic Dancer illustrated a profound sense of ungroundedness that some audiences had experienced when viewing *Earthrise* and other images of the Earth from space. There was an apparent "unsteadiness," particularly in later images of the Earth from lunar orbit, in which the Earth is presented sideways, in the shadow of a looming vertical lunar horizon (Helmreich, 2011, p. 1215). These images, writes Benjamin Lazier (2011, p. 610), "confound one of the presuppositions of phenomenological analysis, that the body has a customary orientation in space: up and down, front and back, above and below, before and behind." It is this ungroundedness, through its association with weightlessness, which has been the focus of recent creative engagements with space (Eshun, 2005, p. 28). This sense of unsteadiness, and the loss of an "upright" orientation, separates the common experience of being grounded on Earth's surface from the apparently liberating experience of space travel that frees the human body from the sensation of being earthbound.

Images of the Earth from space have also prompted audiences to appreciate that the ground to which we are bound is not representative of the whole Earth. These images of a blue planet revealed not "Earth as ground," but "Earth as sea" (Helmreich, 2011, p. 1216). Satellite imagery of the past few decades has captured the gradual spread of blue across the Earth's surface, revealing the rate of change of the melting polar ice caps. The increasing ubiquity of these blue-Earth images has, argued Davide Rapp (2014, p. 2), influenced our approach to design. Rapp suggests that images of the Earth from space are responsible for a trend away from green and toward blue as a signifier of sustainability and ethical design. While "green" was once a synonym for "sustainable," images of the Earth from space have revealed that the man-made golf courses and synthetic lawns of Las Vegas appear greener than the forests of Brazil. Rapp argues that we now recognize green as an artificial camouflage, employed by manufacturers to mask the artificiality of their wares.

Aside from the view of Earth from space, weightlessness offers the most enticing extraterrestrial experience for potential space tourists and creative practitioners alike (Peeters, 2010, p. 1627). The appeal of weightlessness for artists and designers lies in the potential to liberate their practice from the laws that govern the behavior of earthbound objects. Opportunities for creative practice in space are, however, limited, leading practitioners to seek alternative opportunities for creative exploration of the weightless environment. Artists and tourists have taken advantage of the space-like environments created as part of governmental space programs, including simulated weightlessness. The weightlessness that is experienced in orbit is not true zero gravity, but rather free fall. An orbiting satellite perpetually falls toward the Earth, as the Earth rotates away from it. The same weightlessness that is experienced in an orbiting space station is replicable in a variety of ways within the Earth's atmosphere. Tall drop chambers such as the European Space Agency's ZARM tower in Bremen, Germany, enable weightlessness testing on a small scale, inside containers that are dropped from a great height. For humans to experience weightlessness in free fall, they must be elevated to an even greater height on board an airplane. In 1950, Fritz and Heinz Haber first proposed the possibility of staging free fall in parabolic flights to produce a weightless effect for astronauts in training (Karmali and Shelhamer, 2008, p. 594). So-called zero-g flights descend at such a steep angle that passengers experience 20–30 seconds of free fall. Aircraft fly in a series of parabolic arcs, so that the force of gravity increases to up to 1.8 times its usual force on the upward slope, and then decreases to 0G (zero-g) as the craft pulls into a downward slope on the other side of the arc. Parabolic flights such as those offered by the Zero Gravity Corporation are, for now, the most common means of experiencing weightlessness. They offer commercial tourists, researchers, and creative practitioners access to a space-like environment in which to experiment, without leaving the Earth's atmosphere. Commercial operators such as the Zero Gravity Corporation in the United States, or Star City Zero Gravity in Moscow, offer 10–15 parabolic maneuvers creating 20–30 seconds of weightlessness each.

The availability of commercial parabolic flights has led to citizen experimentation with weightlessness. Activities carried out on board reduced gravity aircraft range from scientific experiments to performance art. Since 1995, the Arts Catalyst has supported artists, performers, and philosophers in works that engage with the sciences, including a number that attempt to demystify the "private world of astronauts and cosmonauts" (Frenais, 2005, p. 10). On board parabolic flights, these practitioners explore notions of "groundlessness," experiencing the defamiliarization of their previously earthbound practices (Eshun, 2005, p. 28).[8] Funded by Arts Catalyst, and in collaboration with the French National Space Research Centre (CNES), choreographer Kitsou Dubois became the first professional artist to explore weightlessness on board parabolic flights (Frenais, 2005, p. 9). Dubois (2001) describes how the weightless environment of a parabolic flight allows in her performance an "extraordinary fluidity of movement," in which a performer cannot predict the outcome of her movements.[9] The effects of weightlessness are doubly extraordinary in those performances that Dubois has choreographed for multiple dancers. Since there is no center of gravity, each performer must "create subjective egocentric references." Without a common understanding of up or down, performers "build their own structures on [a] subjective axis" (see Figure 1.7).[10]

Visual artists have also made use of parabolic flights to explore how weightlessness challenges the primacy of vertical orientation. Artist Frank Pietronigro (2000, p. 169) took advantage of a NASA/Texas Space Grant Program–funded parabolic flight to explore how painting could be separated from "the structural support [of] the canvas," with the intention of "identify[ing] new practices for paintings in weightlessness." For *Drift* (1998), Pietronigro constructed a vinyl chamber, within which he released paint of varying consistencies, during the weightless phases of his parabolic flight. The artist was able to observe the "ethereal movement" of the wet paint, once released from his selection of bags and containers, as it drifted around the chamber. As a free-floating body in the center of the chamber, he was able to "increase the point of view from which a painting may be viewed, from a finite 180° semi-circle

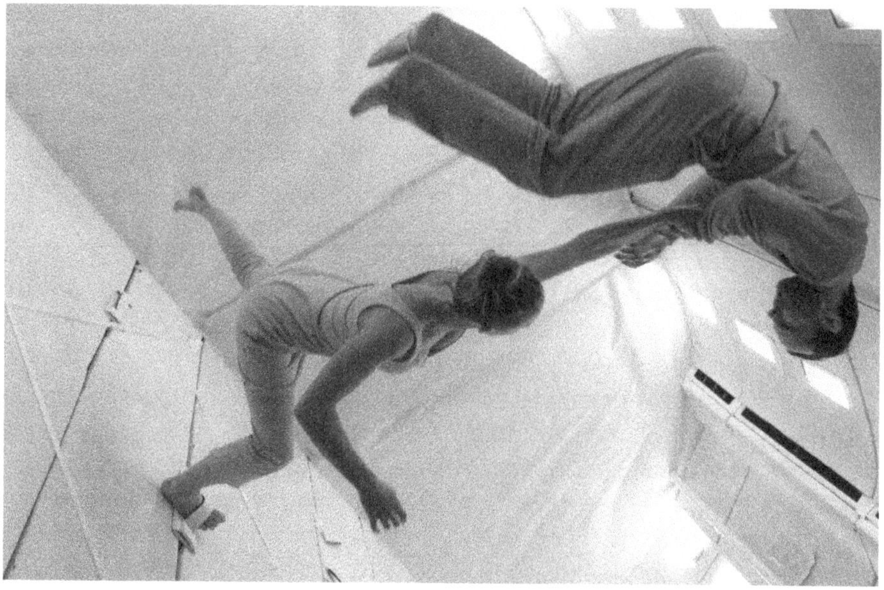

Figure 1.7 Kitsou Dubois, who has choreographed dance performances for weightless environments. Her experience of performing on board parabolic flights has shown her that, for the weightless body, orientation is subjective, and so performers must find "a new system of references" to orient their bodies. She notes that "an upward or downward movement of the eyes" locates objects and surfaces in or out of alignment with "body's longitudinal axis, rather than the gravitational vertical" (Dubois, 1994, p. 60). © Quentin Bertoux

to an infinite number of global viewpoints" (p. 173). His conclusions, along with those of Dubois, foreshadow one of the concerns that may become central to design for microgravity: the need to consider an infinite number of viewpoints (see Chapter 2).

While practitioners lucky enough to find themselves on board a parabolic flight tend to focus on the extent to which its 20–30-second windows of weightlessness replicate the experience of being on board an orbiting space station, few have explicitly considered the variation in gravity that is peculiar to these flights. In 1999, Cosmokinetic Cabinet Noordung staged the first theatre performance in alternating-microgravity conditions on board a Russian parabolic flight, entitled *Gravitation Zero—Noordung Biomechanics* (Grzinic, 2003, p. 82). The performance was observed by an audience of sixteen, inside the craft that had been redesigned as a theatre space. As fellow passengers, the performers and audience experienced the same fluctuations in gravity as the craft flew its ten parabolas. Unlike more recent performances that are contained within the 25-second window of microgravity that is created in each parabola, Cosmokinetic Cabinet Noordung's performance explored the fluctuation of gravity as it alternated between zero-g, 1.8g hypergravity, and the 1G of steady horizontal flight. Director Dragan Živadinov positions the piece as an example of "postgravity art."

The short length of each period of weightlessness experienced on board a reduced gravity aircraft can be seen to limit engagements with weightlessness by creative practitioners. Performances such as those funded by Arts Catalyst must be choreographed precisely, to be completed in less than half a minute. This short time is too little for the production of most forms of visual art, meaning that work must be created on Earth's surface—to be viewed, rather than produced, in the weightless environment. British artist Nasser Azam sought to be the first painter to create work in this environment, for a project entitled *Life in Space*. Azam's *Homage to Francis Bacon: Triptych I* and *II* had to be created almost entirely on the ground before the flight. He then completed the final few strokes on board a parabolic flight from Russia's Star City in July 2008. The microgravity environment forced him to switch from his preferred medium of acrylic paint to oil pastels in order to prevent floating pools of paint (Collett-White, 2009).[11]

Artists and designers now recognize that the microgravity environment demands a radically new approach to practice. Multimedia artist Eduardo Kac (2005, p. 18) argues that we have a lot to learn from "gravitropism"—the way in which gravity affects growth. On Earth, biological forms are "conditioned by gravity" to behave in predictable ways, with shoots reaching upward and roots reaching down. Seedling growth experiments on board space shuttles have shown that, in microgravity, "growth direction is uncontrolled." Each plant finds its own unique and unpredictable growth pattern (Takahashi, 2003). Arthur Woods (1993, p. 297) describes similar contrasts affecting man's approach to form in gravity and microgravity. He observes that sculpture has evolved as "a terrestrial art form," restrained by the "gravity constant," so that all sculptures have a "resting point" or other "point of contact." When designing forms for a microgravity environment, when the "gravity constant disappears," he suggests that there is a need to revisit the conceptions about "aesthetic rightness" that have been formed in our terrestrial environment (p. 298). We must recognize, proposes Kac (2005, p. 18), the extent to which gravity informs the design of objects that we create on Earth, so that we can understand how objects intended for zero-g must be "radically different."

For designers, that "radically different" understanding of objects and how they behave necessarily results in the reinvention of everyday tools and products that are, on Earth, relatively simple in design. In the emerging field of space architecture, it has been recognized that "gravity is the main ordering principle" in existing man-made design (Doule, 2014a, p. 93). This deference to gravity must be reconsidered in order to design for space. Indeed, principles of upwardness and downwardness, or the notion that an object should have a bottom or a top, might ultimately become outdated "relics of humans' bounds to the Earth"

(p. 93). In space architecture, the new "structural geometry is driven by maximum efficiency regarding internal overpressure relative to external vacuum, resulting in spherical, cylindrical and toroidal shapes." Aspects of design that are familiar on Earth become redundant, including bases, feet, or other surfaces designed to create points of contact between the ground and a designed object.

Everyday objects whose fundamental characteristics have been fixed for hundreds if not thousands of years, such as drinking vessels or eating utensils, have been found not to function in microgravity. Eating and drinking in space can be experienced as uncanny distortions of the same activities on Earth, with familiar substances and utensils that behave in unfamiliar ways. Annalisa Dominoni (2015, loc. 126) uses the example of a spoon, a tool that is near ubiquitous on Earth, and whose design is so perfect that it has been unaltered for thousands of years, but that fails to serve its primary purpose of holding or transporting food in microgravity. Similarly, the coffee cup is rendered useless in space as liquids cannot be poured. When contained within a vessel, liquid will cling to the vessel's interior surface, regardless of the vessel's orientation (Pettit et al., 2011). A small jolt will release the liquid from the vessel, so that it floats free, becoming an uncontrollable "gelatinous sphere" (Gorman, 2015). These effects render it virtually impossible to drink from a cylindrical cup or glass. Astronauts typically make do with sealed pouches, from which drink is sucked through a straw.[12] In an effort to resolve the problem, the zero-gravity coffee cup was developed by physicist Mark Weislogel and astronaut Don Pettit, along with mathematicians Paul Concus and Robert Finns who had performed previous analyses of "capillary flow," a phenomenon by which liquids can be channeled along the edge of a container in which two sides meet at a sharp angle (Concus and Finn, 1969). Pettit et al.'s (2011) patent acknowledges not only the practical value of a coffee cup that doesn't spill its contents, but also the social and psychological value of a cup that enables toasting rituals, and that allows use "consistent with how beverage cups are used on Earth."

The ritual of drinking as we do on Earth is made possible for space tourists by products such as the Zero Gravity Cocktail Glass, developed by the Cosmic Lifestyle Corporation, and the Space Glass, developed by Open Space Agency's James Parr for whisky distiller Ballantine (2015b). Both are 3D printed products that could be manufactured on board the ISS on demand, and demonstrate a burgeoning interest to transport Earth culture beyond Earth's atmosphere. COO of the Cosmic Lifestyle Corporation, Samuel Coniglio, identifies the company's remit as the creation of "domestic products for space" (Gorman, 2015). Coniglio stresses that, when designing for space, it is necessary to "start from scratch" (Gorman, 2015), to abandon preconceived notions of some of designs' most fundamental principles. In developing a cocktail glass for use in microgravity, Coniglio and his team of designers aimed to maintain the familiar silhouette of a martini glass, while incorporating channels that would allow it to draw liquid toward the mouth of the drinker, and hence to replicate the experience of drinking in familiar, Earth-gravity conditions. Faced with the redundancy of a flat base, the team chose to create a stem for the glass that terminates in a ball instead of a foot.

The rush to prototype products for use by space tourists reflects not only a desire to innovate, but also a drive to enable the performance of Earth customs and rituals outside of Earth's gravity. Ballantine brand director, Peter Moore, explains that Ballantine's development of the Space Glass had "the ritual of the whisky drinking experience at its heart" (Ballantine, 2015b). Their aim was not merely to make whisky drinking possible in microgravity, but also to transport whisky culture beyond Earth's atmosphere. As humans travel further into space, they will carry with them artefacts of human design. Colonization will be as much a social and cultural process as it will be a technical achievement. Space explorers will make the first steps toward populating our solar system with human innovations and cultures, including clothing cultures.

The fledgling field of product design for space has so far focused on products for luxury lifestyles. Peter Dickens (2009, p. 71) identifies space tourism as the height of conspicuous consumption, thereby aligning it with the many fashion practices that have been declared similarly wasteful since the publication of Thorstein Veblen's (1899) *Theory of the Leisure Class*. Both Ballantine and the Cosmic Lifestyle Corporation position their products in response to a model of space tourism that "emphasizes the concept of luxury, and the lifestyle of hyper-consumerism" (Billings, 2006, p. 162). Engineers and designers have anticipated a need for products on board orbital hotels, where they anticipate that users will carry out leisure activities similar to Earth tourists. This reframing of space as a luxury destination has led to design solutions that enable leisure activities in weightlessness, and in particular social activities such as drinking.

As luxury fashion on Earth has shown, catering to the luxury market does not necessarily mean a requirement to use luxury materials, and high fashion can achieve its status regardless of the materials used in its construction. The designer's identity and the context of a garment's creation can greatly affect its perceived value. Consider, for example, Martin Margiela's *Artisanal* plastic bag and gaffer tape gown (2008), constructed from "raw materials [that] are blatantly incongruous with this [luxury] status," elevated to that status by the time and labor invested in their construction, the designer's reputation, and the restricted availability of each piece in the collection (Groom, 2011, p. 505).[13] Garments designed specifically for microgravity become exclusive by having features that are inextricably linked to exclusive environments and experiences. The spectacularly unfamiliar behavior of everyday materials in microgravity transforms them to such an extent that, through use in a weightless environment, they attain exclusivity.

Exclusivity plays a significant part in the perception that a garment is a luxury product. Space travel is, and will continue to be for some time, an exclusive activity, but this fact alone is not enough to automatically make any product associated with space travel a luxury item. There are numerous items currently in use on board the ISS that are, arguably, ordinary and uninteresting, ranging from ziplock bags to duct tape. This is partly because such items were established for use on Earth before they were adopted for use in space and are, in many cases, available on the mass market. The routine wear of NASA astronauts, for example, is available off the shelf (see Chapter 3). Future space fashion may be made exclusive through limited availability (perhaps available only to spaceship passengers), or as a result of design features that are directly affected by the exclusive space travel environment. It follows that, in order to appeal to the luxury market, spacewear should feature components that function as intended by the designer only in the weightless environment. Such elements may be, for example, shapes that are maintained only in microgravity, or that in some other way provide visible evidence of weightlessness. It is interesting to note that the form-fitting or skintight silhouettes of Space Age catsuits do not fulfill this criterion, as such designs take the same form in space as they do on Earth.

Faced with the reality of space tourism, the fashion industry now needs to move away from symbolic and aspirational references to space travel (such as those communicated in silver or white form-fitting suits), and toward an approach that considers the challenges of dressing a weightless body. Space Age fashion presented a mishmash of allusions to space travel, which were effective at symbolizing Man's conquest of space but impractical in a real microgravity environment. As spacewear emerges as a new field of fashion, it can be distinguished from Space Age or futuristic fashion by its concerns for the realities of space travel. Spacewear must consider the effects of weightlessness on fabric and fit, and must accommodate the malformation of the body that occurs in weightlessness (see Chapter 4). Spacewear must be designed with an understanding of the conditions of the extraterrestrial environments in which it will be worn. Since the conditions of space travel environments vary, most significantly between intravehicular and extravehicular locations, spacewear designers must begin with an appreciation for the particular requirements of the location where their garments will be worn.

While familiar rituals and experiences play an important role, it is essential that these new products do not entirely replicate what happens on Earth. Since the appeal of space tourism lies partly in the novelty of experiencing weightlessness, space products must accommodate and emphasize, not mask or erase, the strange behaviors of familiar objects and materials. Just as tourists choose to wear and be photographed in a vacation wardrobe when they travel away from home on Earth, tourists in space will "want to put something on that signifies that situation" (Mark Timmins, personal communication). In conversation with the author, Susan Buckle of the UK Space Agency describes how providers of parabolic flights recognize their passengers' desire for photographs that record visible evidence of the effects of weightlessness. Buckle has experienced weightlessness on numerous occasions with the European Space Agency and UK Space Agency, and is familiar with the sensations generated by short-term weightlessness. Buckle describes how even the relatively form-fitting flight suits that are required on board the flights that she has flown with Novespace, France, are lifted away from the body. It is not only the body that feels weightless, but also its various parts, and everything that is in contact with it. Her long hair, for example, is lifted outward so that it no longer weighs on the scalp. Buckle recalls an early flight during which her instructor encouraged her to "let [her] hair loose," recognizing that it would allow for more interesting photographs in which her weightlessness is visibly evident.

Her many experiences of weightlessness and work alongside astronaut Tim Peake have shown Buckle that part of the pleasure of space travel is in the defamiliarization of familiar experiences. Buckle describes the importance of visible evidence of weightlessness, not only for participants of parabolic flights, but also for seasoned astronauts journeying to space. Traveling to the ISS on board the *Soyuz* rocket, astronauts have been known to have a "mascot," often a child's favorite toy, "dangling from the roof of the rocket." The purpose of the toy is to evidence the moment when the contents of the cabin becomes weightless, as it is the only moveable object that is not strapped firmly in place. This visible evidence of weightlessness is cause for celebration, signifying not only the success of an individual flight, but also, more broadly, the continued progress made toward conquering the final frontier. "As soon as they get into weightlessness that toy will start floating," reports Buckle, "that's when [the astronauts] all cheer and know they are in microgravity." Buckle's observations about the desire among spaceflight and parabolic flight participants to visibly evidence weightlessness suggest that future space tourists will seek spacewear that, unlike many of the skin-tight or form-fitting fashions of the Space Age, will extend away from the body in ways that celebrate the visible effects of weightlessness, through changes in silhouette and the separation of garment from body.

Defamiliarization of everyday experiences is at the heart of James Parr and Samuel Coniglio's redesign of everyday objects for microgravity, and spacewear designers could learn from their approach. When James Parr was set the task of redesigning a whisky glass for space, he began by recognizing that some substances and materials cannot, and should not, be controlled in space as on Earth. Parr's solution was not to attempt to "control where the liquid goes" but rather to control "the vessel around the liquid"; "Rather than trying to move the whisky, we moved the glass around the whisky instead" (Ballantine 2015a). This kind of inverted approach may also suit spacewear design. Materials behave so differently in microgravity that some of our fundamental assumptions about product design need to be discarded. While on Earth we might constrain or contain materials in order to maintain control, for space travel designers may choose to design in ways that permit the fascinating and entertaining behaviors that materials naturally exhibit in the weightless environment.

Existing experimental fashion for space often fails to account for the extent to which materials behave so differently in a microgravity environment. In 2006, the University of Tokyo hosted finalists from the *Hyper Space Couture Contest*, a spacewear design contest that challenged designers to develop garments for travel in weightless conditions. The contest was partly funded by Rocketplane Global, a commercial spaceflight operator who, at the time of the contest, planned suborbital commercial flights. Chuck Lauer, vice president of Business Development for Rocketplane, describes the company's decision not to restrict its customers' wardrobes, telling Leonard David (2006) of Space.com that "Rocketplane made the fundamental design decision to fly in a true 'shirtsleeve environment' very early in the development program," feeling that free clothing choices would "enhance the customer's space flight experience." They planned to ultimately release a catalogue of garments, making the collection commercially available to space tourists. Rocketplane and its partners recognized the need to offer spaceflight participants a choice of wardrobes, but did not direct competition entrants to consider the effects of weightlessness. As a result, only two of the shortlisted entrants, Asami Okutou and Hitomi Kudou, directly considered weightlessness in the development of their garments.

The ten *Hyper Space Couture* finalists demonstrated in their designs a variety of approaches to spacewear, arising from a range of aesthetic and practical considerations other than weightlessness. Despite the contest's relationship with Rocketplane, many of the shortlisted designs demonstrate an approach similar to that taken by Space Age designers, as in Shiori Matsuda's rigid, disk-like neckpiece, and Akie Masuda's asymmetrical miniskirt, both of which overlook the practical limitations of their designs and instead focus on aesthetics and signification. The designs do incorporate some visual references to newer technologies, but are ultimately limited by the same prioritization of symbolism over function. Most make a token reference to the practical demands of a spacesuit, without accommodating the technology that would allow them to function in a real extraterrestrial environment. Okutou and Kudou, two of eleven finalists, succeeded in directly responding to the effects of weightlessness. Asami Okutou's floor-length gowns prioritize elegance and spectacle over practicality. Loose fabric strips are designed to float free when weightless, swirling around the body as petal-like appendages. The design embraces the effects of microgravity, exploiting weightlessness to create new, fluid silhouettes. Okutou's design resembles a "typical kimono" in Earth gravity, but weightlessness causes its "dozens of black, ribbon-like layers on its surface flutter, showing glimpses of the colorful reverse side" (Katayama, 2007). By contrast, Hitomi Kudou's response is to try to erase the effects of weightlessness, recognizing the lack of control over silhouette that result from loose fabric, by designing a skin-tight, white jumpsuit.

A more recent collaboration between the European Space Agency and five European fashion schools—*Couture in Orbit*—reveals that a lack of consideration for the effects of weightlessness is still evident in speculative spacewear design. This may be in part because each of the five schools involved in the project in 2016 were given themes on which to focus, none of which directly related to weightlessness or variable gravity. At the Politecnico di Milano, led by Annalisa Dominoni (2016), students were asked to focus on connections between life in space and on Earth, considering both technological and emotional aspects of design. Like the finalists of the *Hyper Space Couture Contest*, Dominoni's students were "inspired by space images" to select "colors like silver and white . . . materials such as aluminum and reflective surfaces, selecting curved shapes." They found particular inspiration in images of astronauts' postures and movements in the microgravity environment of the ISS. At the ESMOD Berlin International University of Art for Fashion, students under Philippe Ara (2016) found inspirations in the patterns and textures in images of the Earth viewed from space. They studied "sports and body monitoring technology," and considered the use of "garments as protection in dangerous environments" (Science Museum, 2016). Students of Ravensbourne in London were lucky enough to visit the collection of spacesuits at the Science Museum, and as a result took a more technical and scientific approach to the brief, and were asked to consider the possible colonization of alien planets. Inspired by British astronaut Tim Peake's military past, many of the Ravensbourne students incorporated elements of parachutes into their designs (Selina Pang, personal communication). In Copenhagen's Fashion Design Akademiet, students were invited to consider the future of everyday commuting in space (Science Museum, 2016). Each group formed a relationship with established commercial suppliers to the European Space Agency, who presented a range of specialist textiles and wearable technology to inspire the students in the development of their designs. The resulting designs ranged from whimsical responses to the Space Age aesthetic to more technically feasible designs that incorporated technologies provided by the commercial sponsors.

One Japanese designer, Eri Matsui, has succeeded in responding to the problem of designing garments for a microgravity environment, and has demonstrated the effectiveness of his design on board a parabolic flight. Parabolic flights have yet to provide a venue for a fashion event, but have hosted the wearing of Matsui's wedding dress. The Zero Gravity Corporation typically provides its passengers with regulation flight suits, emblazoned with the company's logo. However, they do make exceptions, as was the case for the wedding of Erin Finnegan and Noah Fulmor, who held their ceremony on board a zero-g flight in 2009. For the weightless ceremony, guests wore suits and ties by J Lucas Clothiers while the bride wore Matsui's dress. Matsui had developed her zero-gravity wedding dress for brides who wish to wed while weightless, making it an ideal choice for Finnegan. Matsui's dress is a compromise between style and function, working both with and against the weightless conditions for which it was designed. It features waterfall ruffles, running vertically from shoulder to waist and below the waistline, columns of draped silk that appear to form a floor-length skirt. This surface layer makes the most of microgravity's ability to lift draped cloth, "designed to float like a sea anemone in the ocean" (Finnegan, 2014). Beneath is a much more functional layer of separate bodice and trousers, designed with functionality in mind. Matsui recognized that a loose skirt attached at the waist would not drape downward without gravity, potentially exposing the entire lower half of the wearer's body.[14] In the weightlessness of a zero-g flight, the form-fitting lower layer of Matsui's dress secures the garment in place, ensuring that the body remains reliably concealed while only the outer layer floats free, and prevents the immodest billowing of the skirt that might otherwise result from weightlessness.

Though largely speculative, the few projects described above reflect a curiosity about the potential need for fashion design for space, and provide evidence of the aesthetic possibilities of such fashion. They also, however, demonstrate that designers are still uncertain about how to approach this new frontier and, in many cases, cling to the familiarity of the Space Age aesthetic. Asami Okutou's design aside (and forgiving Matsui's use of white for a wedding dress), these existing examples of spacewear reveal an ongoing dependence on Space Age symbolism that is perhaps best demonstrated in designers' color palettes. The dominant colors in the shortlisted designs for *Hyper Space Couture* and *Couture in Orbit* are white, silver, black, and blue. These are, observes Arthur Woods (1994, p. 4), also the most dominant colors on board a space station, largely because the view from a space station's portals consists only of the blackness of space and the white and blue of the distant Earth. These also continue to be the colors that dominate images published by NASA.

One could argue for the logic of symbolizing space travel through the use of those same Space Age colors. Arthur Woods, however, proposed that his own *Cosmic Dancer* sculpture, to be installed on board Mir, should not blend into its background but, rather, should provide contrast in the otherwise "drab" interior of the space station. Woods ultimately selected green for his kinetic sculpture. In making this choice, he took a significant step away from the conventions of Space Age design, recognizing that once it has been relocated to space, art and design moves beyond the need to symbolize space travel. Instead, he identifies the emergence of a need to symbolize Earth, and to connect space travelers with their terrestrial home. Spacewear designers have yet to reflect on their dependence on the Space Age aesthetic. Many of the designs that were developed for the *Hyper Space Couture Contest* and *Couture in Orbit* are naïve, informed by aspirational imagery and more concerned with the idea of space travel than the practical reality, but following Woods's lead, it is necessary to take a more critical approach to those aspects of Space Age design that are still taken for granted when developing new artefacts for space.

These criticisms aside, it is also important to recognize the value of spacewear contests in their pioneering efforts to give credibility to a field that was once more closely aligned with science fiction. These contests demonstrate the progress that has been made since the Space Age, in terms of directly addressing the practical and technical requirements of fashion for space travel. All of the colleges involved in *Couture in Orbit* were given access to ESA space-certified textiles, and encouraged to incorporate these into their designs (Ara, 2016). Thus, there is a move toward producing garments that are in line with the regulations and requirements of real space programs. Where designers have begun to directly engage with weightlessness, they have begun to exploit new opportunities to directly engage with the spaceflight industry, not least through the increasing availability of parabolic flights. Activities that take place on board parabolic flights offer an insight into the kinds of experiences and clothes that are likely to appeal to future space travelers, and how these needs will differ from those of Space Age consumers. Providers of parabolic flights recognize that their customers seek the novelty of weightlessness in ways that can be visibly evidenced, perhaps partly in the wearing of clothes designed to respond to weightlessness.

The potential for mass commercial space travel, and hence for mass occupation of weightless environments, invites the birth of a new market for product designers, interior designers, and, as this book argues, for fashion designers. The projects described above are small steps in the journey toward fashion garments that might actually be worn on board a commercial space destination, but significant nonetheless. While these examples are few, they can be considered in conjunction with a range of fashion practices in more conventional settings. Having understood the characteristics of the microgravity environment and the ways in which they affect the behavior of a garment, as distinct from the aesthetic traditions of Space Age fashion, it is possible to identify existing fashion practices that will inform design for that new environment. As the next chapter will show, there is a lot that can be learnt from fashion that was not intended for space.

2
WEIGHTLESSNESS ON (AND ABOVE) THE CATWALK

Space will introduce new conditions that fashion designers have not previously considered but, nonetheless, there are ways in which spacewear designers can learn from existing fashion practices. There are features of weightlessness, and behaviors of weightless objects, that can be mapped onto everyday phenomena that take place in Earth gravity. Hence, spacewear designers can benefit from observing certain practices that take place on the Earth's surface. Spacewear designers can understand the effects of weightlessness by comparing them with more familiar phenomena, beginning with an understanding of how orbit works. Objects in orbit, such as the International Space Station (ISS), do not experience gravity as they do on the Earth's surface. The gravitational conditions experienced in orbit are known as microgravity. Gravity pulls an orbiting object toward the Earth and, as it falls, the Earth rotates away from it, so that it falls perpetually without ever crashing or landing. In this state of perpetual free fall, objects are not truly weightless; rather, they fall at the same rate as other objects around them, so that they experience what appears to be weightlessness (Karmali and Shelhamer, 2008, p. 594). Crew on board a space station fall at the same rate as the station, remaining constantly at a distance of approximately 400 km from Earth's surface (NASA, 2011). Once we understand orbit as free fall, we can relate weightlessness to more familiar experiences, such as floating or falling.

These common sensations—floating and falling—are the subject of many fashion practices that do not directly reference weightlessness but that, by relating to associated phenomena, can teach us about ways to design for the microgravity environment. The fashion industry makes use of underwater photography, vertical catwalk shows, and other forms of suspension that introduce many of the same phenomena that are encountered in space. Viewing the effects of these practices on Earth fashion can provide insights into the ways in which garments may behave, or be encountered, in microgravity. When floating or suspended, clothes and the bodies that wear them are inverted or distorted, adopting new forms and draping in unfamiliar ways, thereby exposing features of the garments that are not otherwise evident. These effects can be deliberately accommodated in the process of spacewear design.

Suspension and rigged displays

The idea of weightlessness was desirable in the fashion industry long before Man's first ventures into space. Weightlessness has been synonymous with modernity. Coco Chanel's "weightless wools" enabled women to cast off the shackles of a heavy wardrobe, becoming a crucial expression of liberation and innovation (*Vogue*, 1930, p. 108). Before the Space Age, the world experienced the Jet Age. Lightweight clothing was an air travel essential. The need to keep the weight of luggage to a minimum prompted an association between lightweight clothing and modern, luxury lifestyles. The use of the term "weightless" in fashion magazines of the Jet Age suggests a relationship between the desirability of weightlessness and air travel. Writing for *Vogue* in 1947, Mary Jean Kempner (p. 146) explained to readers that air travel meant "editing a wardrobe down to the airlines' sixty-six pounds allowance," and advised packing "weightless" ballet slippers as a substitute for heavier footwear. In 1948, Eleanor Montgomery (p. 135) marveled that, for her sixty-day round-the-globe trip, she was able to fit a complete "weightless" wardrobe of hats, coats, dresses, shoes, and accessories, within her luggage allowance.[1]

As the Space Age approached, the fashion industry saw the invention of new, lightweight fabrics such as nylon and Lycra. These man-made materials represented the peak of textile innovation, and their light weight was presented as a key benefit. The October 1957 issue of *Vogue* describes a mohair and nylon coat as "weightless as a moth's wing." In 1960, a *Vogue* fashion spread introduced Du Pont's new miracle fabric, Lycra (Spandex), with an article titled "The New Thread of the Weightless Matter." The article was accompanied by an image of a Lycra girdle, strung below a helium-filled balloon, soaring above the Manhattan skyline. On the following page, a bundle of similar girdles is shown piled in a weighing scale, accompanied by the text, "Heaped high with all nine of the girdles and brassieres from these and the two foregoing pages, [the scale] registers just 1 3/4 lbs" (*Vogue*, March 1960, pp. 137–139). These images invite women to imagine themselves liberated from the heavyweight undergarments of previous decades, and to consign heavy fabrics to the past. The fashion for describing lightweight garments as "weightless" continued long after designers had moved on from the Space Age aesthetic, with *Vogue* continuing to laud "weightless" fabrics through the following decades and into the twenty-first century (*Vogue*, September 2008, p. 689).

As textiles become lighter and lighter, each era seems to introduce fashions that are more "weightless" than the last, but true weightlessness is an aspiration that fashion designers can never achieve on Earth's surface. With images showing the effects of zero-g beamed back to Earth from the ISS, designers are now able to understand weightlessness in new ways. Designers can now access still images and videos that show how objects and materials truly behave when not weighed down by gravity, not because those objects' density changes but because they exist in a weightless environment. Observing these documents could prompt the fashion industry to revise its understanding of weightlessness, so that it is understood not as a feature of a garment but as a condition that is enabled by the surrounding environment. With access to an increasing array of images of objects in space, in both reality and fiction, some designers have shifted their focus from the pursuit of weightless garments to the display of garments in ways that emulate the appearance of true weightlessness.

Nothing connotes weightlessness more than a garment that is held aloft with apparent ease. Just as *Vogue* used balloons to evidence the light weight of Lycra, contemporary designers have strung their garments from an array of floating and flying objects as if to imply the possibility that they might be spirited away, or that they could spontaneously take flight. Balloons have the advantage of connoting both weightlessness and playfulness, subtly transporting the garment into a make-believe land of flying dresses. Hussein Chalayan's second commercial collection (*Temporary Interference*, Spring/Summer 1995) features garments suspended from helium-filled balloons. The ankle-length slip dresses are carried on fine straps that, instead of looping over the shoulders to trap the body inside the garment, extend directly upward until they connect to the balloons. Some of the dresses have two straps and hence only two balloons, while in others, multiple straps crisscross the neckline, reaching upward to a bunch of balloons that forms a primary-colored cloud above the model's head. Although each of these dresses covers a body, it is lifted so that no weight falls on the model's shoulders. The balloons and dresses are perfectly balanced, with the helium providing just enough lift to keep the dress aloft, and gravity providing just enough weight to keep it from soaring away. The balloons transform the presentation of these otherwise earthly dresses, so that it diverges from the conventions of fashion and into the realm of philosophy and storytelling. For Chalayan, the balloons, seeking to float skyward but tethered to an earthbound body, represent Man's attempts to reach the heavens, and our inevitable failure to do so.

In Chalayan's experiments with balloons, there is not only the implication of a weightless garment, but also consideration for the space that is created between the body and the garment when a garment is lifted from it. For Fall/Winter 2003, Chalayan presented *Kinship Journeys* to audiences in a show that opens with a model on a trampoline, dressed in a black slip that is held aloft by two spherical helium-filled balloons. The show features several further balloons, each tethered to a model either by straps or a harness, making reference to parachutes. The balloons are enclosed in silk tulle, knotted to the strings that tether it to each wearer, in arrangements that reflect the netting depicted in Victorian illustrations of hot air balloons.[2] One model wears a tulle veil that is draped to resemble a collapsed parachute, with its strings trailing limply down the model's thighs. Nicola De Main, who worked with Chalayan on the development and staging of the show, described to me the process of stabilizing the balloons, with the aim of achieving balance between the body, balloon, and garment. Each balloon was required to float serenely above the model's head, floating as if its stability was unaffected by the model's movement as she strides across the stage, and at a constant height so that the garment was neither pulled off nor allowed to fall. The trampoline was included alongside other props as a result of experiments carried out in Chalayan's studio. Chalayan, De Main recalls, "was fascinated by the idea of the dress being held up." "He wanted to know what would happen to [the dress] as [the model] jumped up and down," and imagined the body rising upward to fill the negative space that existed between the elongated straps of the slip and the strings to which they were tethered.

Chalayan (2015) denies a direct interest in weightlessness, saying that he is more interested in the events that follow. What informs his designs is a concern not for floating, but for falling. He aims to investigate the "eventuality" of falling, by capturing "the tiny moment between weightlessness and the crash": "I'm interested in what weightlessness means at the moment [a dressed body] is dropped." This interest is particularly evident in his work outside of fashion. Balloons play an important role in the promotional materials for Chalayan's first theatrical work, a contemporary dance piece, *Gravity Fatigue*, which premiered in Sadler's Wells (London) in 2015. Discussing the show, he describes how he seeks to preserve that "tiny moment" of weightlessness as if his performers and their costumes are in perpetual free fall, much like an object in orbit. As his models or performers jump on a trampoline, or leap across a stage, the clothed body is, for a tiny instant, weightless.

This moment of weightlessness before the fall has fascinated artists, photographers, and designers for many hundreds of years. Art of the late eighteenth century shows some understanding of the effects of weightlessness on fabric that seems to magically transform clothes as they hover momentarily in the liminal phase between rising and falling. Giandomenico Tiepolo's ceiling fresco, *Punchinelli on a Swing* (1797), and Jean Honore Fragonard's *The Swing* (1767), depict clothed bodies on rope swings, in that brief moment of weightlessness between the upswing and downswing. In both of these images, hemlines are shown flaring outward, and ankles are exposed as fabric balloons around the body. Joanna Fassl (2012, p. 143) describes seeing, in Tiepolo's fresco, the sensation that something is "wrong." The rope, which is normally taut under the weight of a body, is loose, as if Punchinelli exerts no downward force on it. Indeed, she acknowledges, he is weightless. After the invention of photography, moments of free fall could be captured and preserved. In Yves Klein's 1960 photograph *Leap into the Void*, the artist is shown leaping from a suburban rooftop. Dressed in a suit, which balloons around his upper torso, the male figure is "simultaneously floating and falling" (Kolodziej, 2012, p. 293).

Contemporary images of this "moment where levity and gravity meet" (Kolodziej, 2012, p. 293) can provide spacewear designers with some insight into how loose-fitting garments behave when weightless. A moment of weightlessness, as a model leaps directly upward, is captured by photographer Arthur Elgort in a photoshoot for *Vogue* (see Figure 2.1). The model, Audrey Marnay, jumps on a bed in a lavishly decorated Parisian hotel room. She is frozen in the moment between rising and falling, at which her acceleration cancels out the force of gravity. Her long chiffon dress no longer drapes downward under the force of gravity, but instead billows outward, distorting her silhouette. For a short moment, the model and her clothes are neither pushed upward by the momentum of her jump, nor pulled downward by the force of gravity, and it is possible to imagine what this dress might look like if worn in microgravity.

Unfortunately for spacewear designers, this moment of weightlessness lasts for only a moment, and cannot be sustained over a longer period on Earth's surface. Designers who desire to recreate the appearance of weightlessness on the catwalk must resort to alternative methods, such as rigging all or part of a garment so that it does not drape directly downward from the body. For their Fall/Winter 2007 show, Viktor & Rolf fitted bodies with rigging that supported a personal lighting and sound system, and allowed garments to be "variously suspended, draped and gathered" (Evans and Frankel, 2008, p. 196). The rigged garments are subject to several different forces acting against each other, being simultaneously pulled upward by the rigging, and downward by gravity. Suspended in this way, a garment exposes the full, unfurled diameter of its hemline, as it might do if unfurled by the conditions of microgravity (see Figure 2.2). Unlike in microgravity, however, the weight of the fabric is evident, as it drapes downward from the points of support at the rigging and the waist. In this way, the attempt to lift the garment upward serves to evidence the weight that is pulling it downward toward the Earth.

In microgravity, both the clothes and the body are suspended in a void, "ungrounded" (D'Aloia, 2012, p. 224). Representations of a void, and the body ungrounded within it, have been produced in two very different ways by Alexander McQueen in *Savage Beauty* (Fall/Winter 2006), and Iris van Herpen in *Biopiracy* (Fall/Winter 2014). McQueen's *Savage Beauty* was celebrated for its holographic projection of Kate Moss floating ethereally inside a glass pyramid (Gleason, 2012, p. 152, see Figure 2.3). For the show's finale, McQueen resurrected a theatrical illusion known as "Pepper's Ghost," conceived by Henry

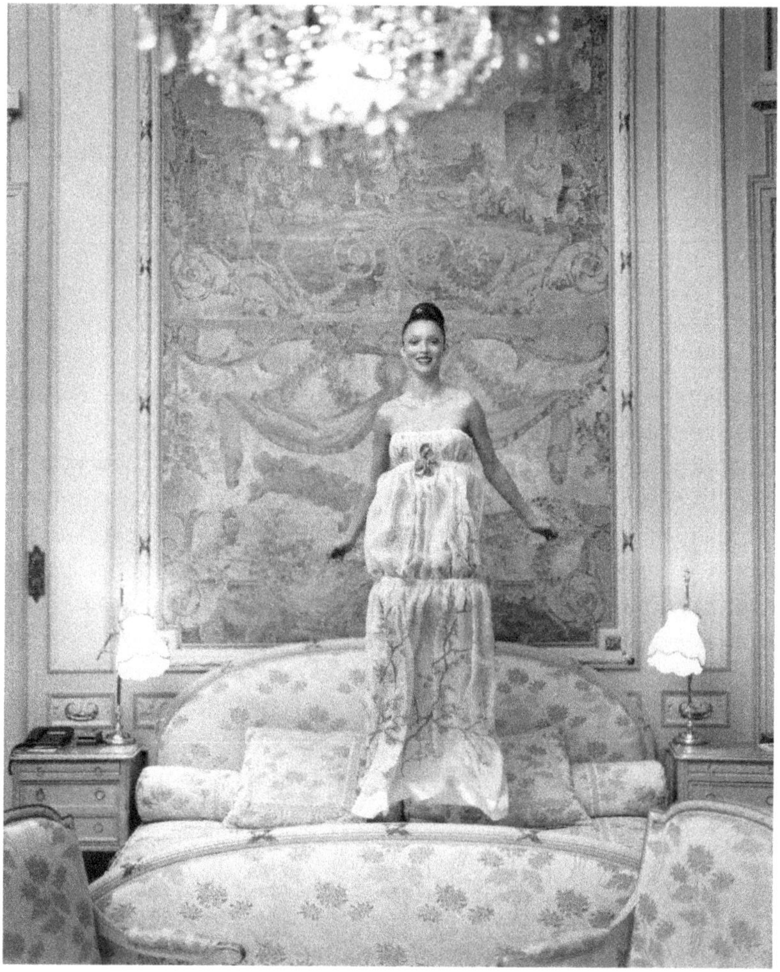

Figure 2.1 Model Audrey Marnay photographed for Vogue, 1999, in a chiffon flower-print strapless dress by Emmanuel Ungaro Haute Couture. The dress billows during the moment of weightlessness that is experienced between rising and falling. © Arthur Elgort/ Conde Nast via Getty Images

Dicks during the industrial revolution and perfected by John Henry Pepper for Victorian demonstrations of the "wonders of optical science" (Ruffles, 2004, p. 29). McQueen's version of the illusion appears to show a specter suspended in a void in the center of a large glass pyramid on the catwalk. A ghostly glow emerges from the darkness in the center of the pyramid, and gradually comes into focus, first as a shadowy silhouette shrouded in floating fabric, and then as a recognizable apparition of Kate Moss in a long gown. The gown undulates around her body, and Moss leans into the swirling fabric as if there is no gravity to

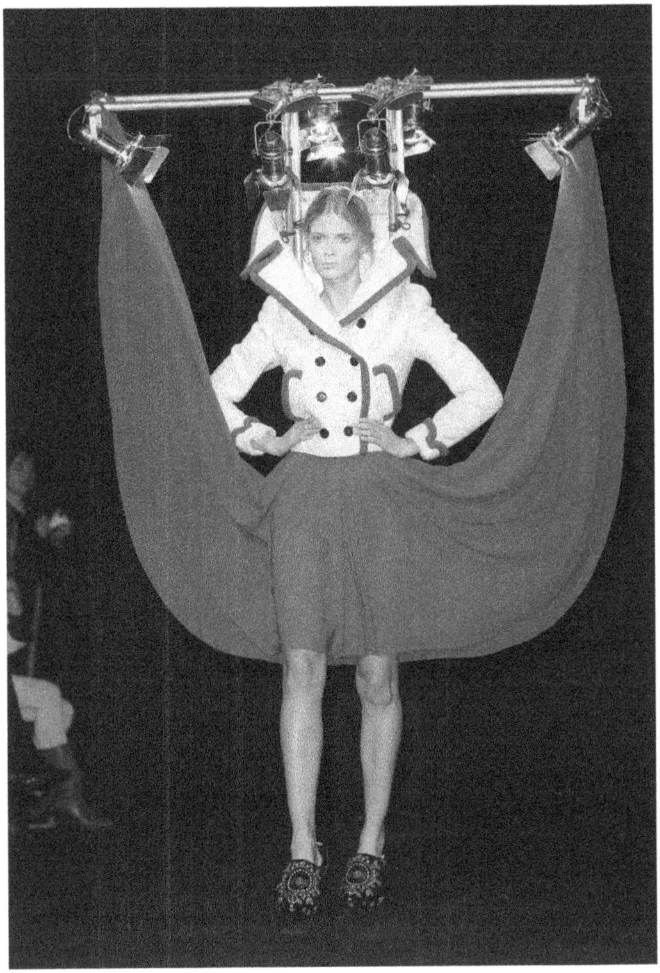

Figure 2.2 Viktor & Rolf Fall/Winter 2007. Viktor & Rolf rigged garments to scaffolding, along with personal lighting and sound systems for each model. The rigging unfurls the hem of the skirt, as if weightless, but at the same time the drape evidences the weight of the fabric as gravity pulls it toward the Earth. © Michelle Leung/WireImage

shake her off balance. Both Moss and her gown appear to be floating, unaffected by gravity.

Other designers have made reference to aspects of deep space travel other than, or in addition to, weightlessness. Iris van Herpen uses vacuum packing to reproduce the appearance of weightlessness in her runway show for *Biopiracy*. Van Herpen's models are suspended between clear plastic membranes, held in place by a vacuum as the air is sucked out from the space between the plastic (Figure 2.4). The membranes, and the bodies that they contain, are rigged vertically

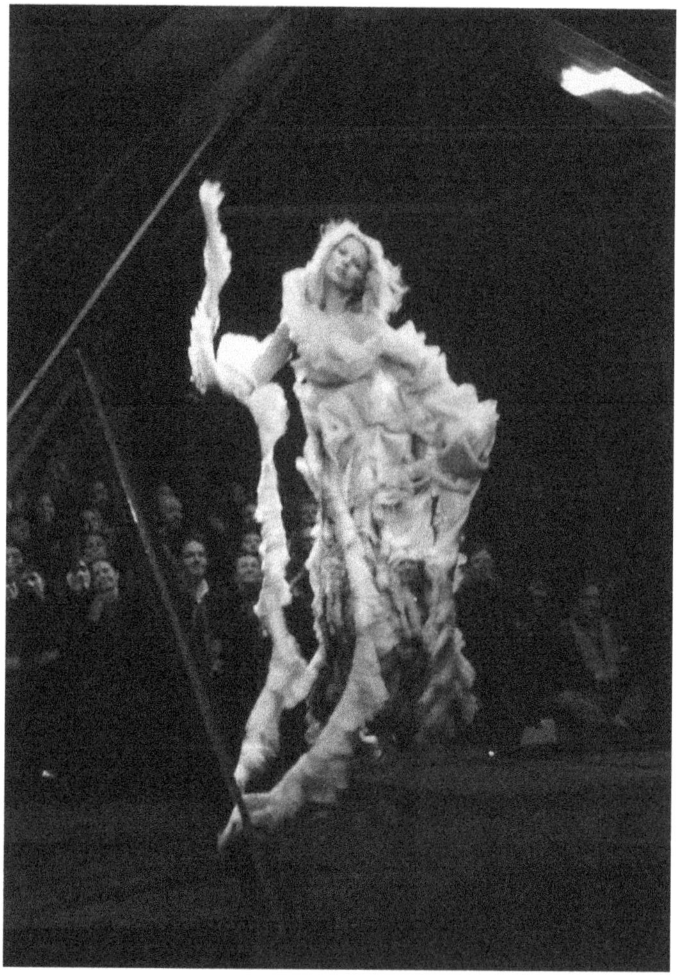

Figure 2.3 Kate Moss appears to hover weightlessly in a hologram at Alexander McQueen Autumn/Winter 2006. © Randy Brooke/WireImage

in the center of the catwalk. Throughout the show, the bodies barely move, as if asleep, or resting in the kind of suspended animation that might enable long-distance space travel (van Herpen, 2014). Elsewhere, similar vacuum packing of the body has been presented in visions of future space travel. "Body architect" Lucy McRae (2016) applies a similar system to participants' bodies in her work, *Astronaut Aerobics Institute*, for the 2014 London Design Festival. The project imagines a future of mass commercial space travel in which tourists attend a day spa that prepares the body for space travel. McRae's study responds to discussions with NASA about "the implications zero-gravity has on the body, and how we will need to prepare for the rigors of more permanent weightless living." Extending these ideas

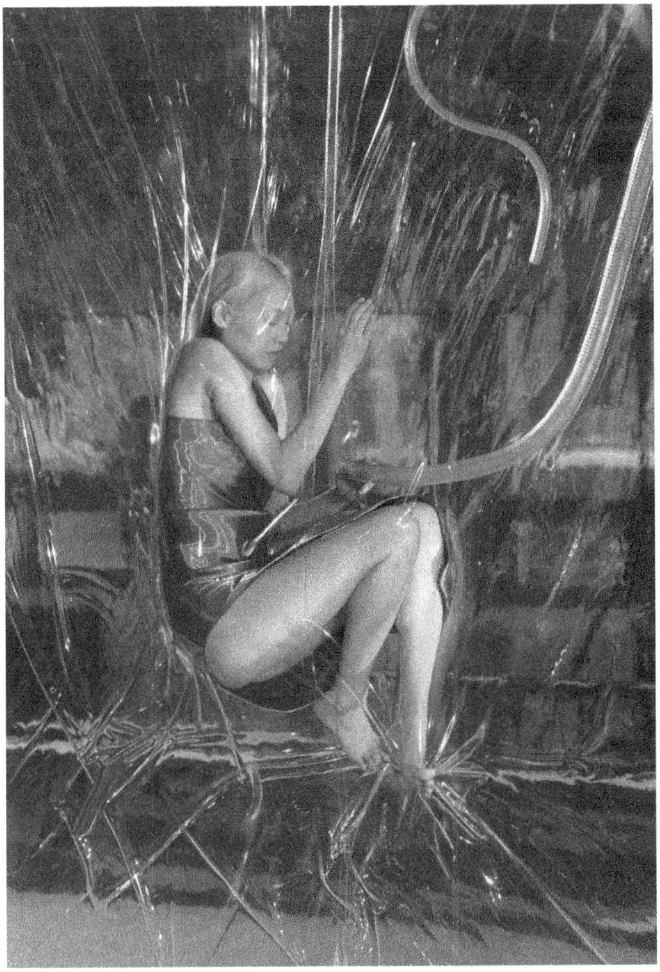

Figure 2.4 Iris van Herpen's runway show for *Biopiracy* (Fall/Winter 2014/15). It features vacuum-packed models, trapped between two plastic membranes, rigged vertically in the center of the catwalk. The models appear to be asleep, or frozen indefinitely in a state of suspended animation. Science-fiction visions of space travel have used images of similar systems to illustrate how the body might be preserved while it is transported long distances in space. © Patrick Kovarik/AFP/Getty Images

into the realm of science fiction, she displays bodies vacuum packed in "aerated cocoons." McRae's early experiments for the project involved stepping between two foil emergency blankets, rigged vertically in a frame, and sucking out the air with a vacuum clearer so that her body is embedded into a wall. She describes the sensation of being "embraced by [a] machine" as the foil hugs the body "from every angle." In these examples, weightlessness is associated with temporal as well as physical suspension, with bodies frozen in time as well as space.

In microgravity, both the body and the clothes are suspended. The lack of gravity acts to make both the body and the clothes weightless, leading to a separation of the body and the clothes that it wears. The body does not form a support for the clothes, nor do the clothes weigh on the body (see Chapter 4). Indeed, the clothes can be considered independently from the body. Weightlessness enables empty garments to come to life. On a planet's surface, under the force of gravity, a soft garment is limp and lifeless without a human body to give it form. This limpness is a result of gravitational force acting upon it, pulling it toward the ground. Without this force, a garment can maintain form even when the body is absent. Microgravity affects the body and its clothes equally, and just as there is a neutral posture for the body, there is also a neutral posture for a garment. It can float as if held aloft by a wearer's body, even when no body is present.

Rigged displays on Earth also enable the separation of clothes and body. The absence of a body within a garment can have the effect of showcasing the form of the garment as if not weighted by gravity onto a human form. Mannequins are used less and less, as clothes are draped over or suspended from abstract objects (Geran, 2006). Issey Miyake often chooses to display his garments in "installations" rather than catwalk shows (Mackrell, 2005, p. 154). His 1997 *Arizona* collection was shown suspended on single wires rather than on models "to emphasize their sculptural abstraction" (Quinn, 2002, p. 150). Japanese artist Lun*na Menoh presented a similar display as installation art. In her 1998 installation entitled *Spring and Summer Collection 1770–1998*, stiffened fabric is displayed as suspended tubes that take the forms of historical costumes (Breward, 2003, p. 15). In each of these examples, there is an implication that the wearability and silhouette of a garment is less of a concern than the qualities of its fabric. Without the presence of the body, the qualities of the garment, not how it is worn, can become the focus of attention.

By separating garment from body, rigging imposes fewer constraints on the presentation of a garment. The designer is free to explore riggings of any shape, size, or height, choosing to emphasize particular aspects or reveal particular details. A high rig can expose the underside of a garment, forcing audiences to look up at it from below. A garment set in motion or suspended high above the ground can explicitly reference the weightless environment of space, as it appears to float high above the ground. In Anya Hindmarch's *Out of this World* floating showcase at London Fashion Week, 2014, orbs painted to resemble planets hang above the catwalk, and pinprick lights are scattered over the black backdrop, creating the appearance of a starry night sky. As models strut the catwalk, Hindmarch's handbags are lowered from the ceiling on invisible wires, and then slowly raised again, as if floating weightlessly.[3] In this setting, the suspended bags appear to become astronomical objects that can be plucked from the stars by passing models, much like a floating astronaut might pluck a passing tool as it floats through cabin of the space station in front of her.

More often, rigged displays in catwalk shows make reference to mythical or fantastical tales, with garments floating as if in ascent toward the heavens. In Gareth Pugh's Spring/Summer 2015 show, *Ascension*, one of several videos was screened instead of a live show to promote the collection. In the slow-motion video, choreographed by Wayne McGregor and directed by filmmaker Andrew Thomas Huang, the ascension takes place so slowly that the models' motion is almost imperceptible, as in a tableau vivant. As the film starts, a group of models huddle on the floor, some draped in strips of white cloth. One model begins to separate from the group, rising upward and, as she ascends, the white cloth is drawn away from the other models, who are left nearly naked in the shadows. It becomes evident as she rises that the draped strips of cloth are attached to the ascending model's white robe, at her shoulder blades. The cloth is, it seems, as weightless as she is. The strips are drawn upward from the tips, and fan outward, taking a winglike form, and transforming the model into an angel-like figure.

Pugh's tableau pays homage to a history of depictions of gravity defiance in the arts. In the visual arts, particularly religious images, "weight and weightlessness" are "visual metaphors for sin and virtue" (Arthur, 2012, p. 52). Giotto's frescoes of Hope and Despair (1305), for example, adorning the walls of the Arena Chapel in Padua, show Hope "floating effortlessly upwards," while the figure of Despair "hangs ponderously, her body so heavy that the rod to which she is attached sags" (Edwards, 2012, p. 76). For more than a thousand years before Newton proposed his theory of gravity, our understanding of the nature of gravity was informed by Aristotle. Aristotle proposed that all things are naturally drawn toward their place of origin, and hence, in Ancient European art and literature, creatures of the heavens are able to fly by harnessing the force that pulls them toward the heavens, "while humans, made from clay, remain . . . Earthbound" (Edwards and Bailey, 2012, p. 2).

Floating fashion

Without gravity, the fabric of a loose garment does not cling to the wearer. Any part of a garment that is not tight or elasticated, or somehow tethered to the body, behaves as if free of its wearer. In the space industry, actions are taken to minimize this effect. Applied pressure is used to force parts of a garment close to the body (Watkins and Dunne, 2015, p. 378). On board NASA's first space station, Skylab, crew were quick to discover the inconvenience of loose clothing in their compact work environment. In order to eliminate the risk of their clothes catching on the equipment that surrounded them on all sides, Skylab crew were dressed in trousers that could be snapped to elastic straps on the hem of the jacket, and that could be kept from migrating up the leg thanks to stirrups (p. 342). While loose, floating fabric may be inconvenient for working astronauts, it may be desirable for a fashionable space tourist. The convenience of a form-fitting suit may be outweighed by a desire to wear a garment that visibly evidences the effects of weightlessness.

Earthbound fashion designers have exhibited an interest in ways of freeing loose parts of a garment from the wearer—from rigging the skirt on a crinoline, to photographing fashion in a windy environment so that fabric is blown away from the body. The underwater environment lifts garments away from the body in ways that, more so than other environments on the Earth's surface, partly replicate the effects of weightlessness. Indeed, NASA has long recognized the similarity of the underwater environment to space, and trains its astronauts underwater in a Neutral Buoyancy Lab, a forty-foot deep tank of water that has been used since 1995 for "suit qualification training" (Gast and Moore, 2011, p. 318). During underwater training, astronauts learn the limitations of their suits, and how they must adapt or "translate" their movement to account for those limitations (see Chapter 4), and get used to the general sensation of wearing it in simulated weightlessness.[4] The underwater environment does not perfectly imitate the weightless environment. As astronauts move in the water, hydrodynamic drag creates force that would not exist in space. Underwater, it requires more force to set an object in motion than it does in true weightlessness, as drag creates an opposing force. For this reason, a dive team must attach weights to the legs of the suit once the astronaut is underwater, in order to more accurately replicate "neutral buoyancy" (Gast and Moore, 2011, p. 321).

Despite the presence of hydrodynamic drag, the underwater environment more closely replicates microgravity than other environment on the Earth's surface, and when true microgravity tests are impractical, water provides a valuable impression of the effects of weightlessness. Fashion photographers have exploited the underwater environment to loosen garments from the body so that they adopt a shape that is independent of the silhouette on which they might otherwise drape. One of the most notable contemporary photographers to explore underwater fashion photography is Zena Holloway. In Holloway's photoshoot for *Cent* magazine, fabric is seen floating as if unaffected by gravity (Figure 2.5). Coats are shown billowing at the hem, and upturned collars reach skyward. The volume created by the ballooning fabric at the chest and hips is offset by a narrow waist, where a button holds the coat in place and preserves a feminine silhouette

The underwater environment illustrates how a garment, when loosened from the body as if by weightlessness, tends to billow outward rather than hang downward. For this reason, length can be transformed into radius. The linear measurement, that one might use to describe the distance that a garment hangs in the direction of the floor, can instead describe the distance that loose fabric might extend away from the body. Fashion photographers have exploited the

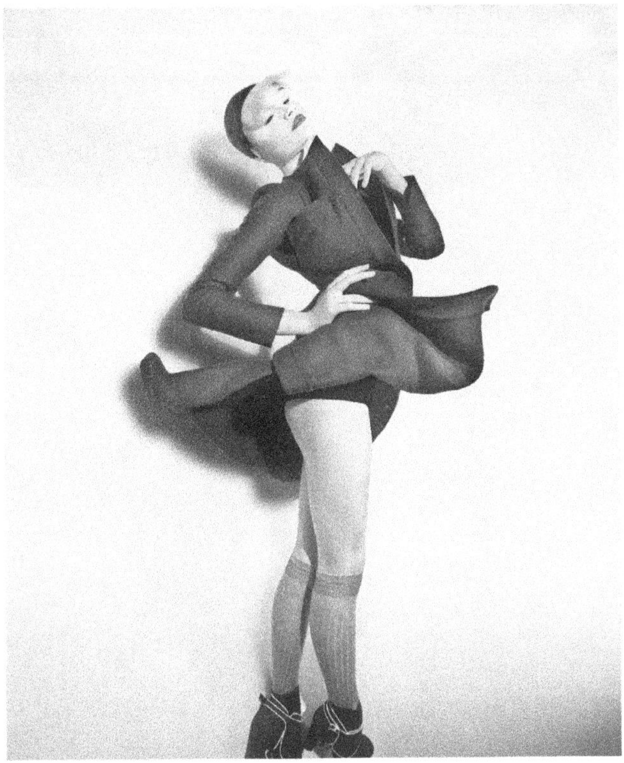

Figure 2.5 An underwater photoshoot by Zena Holloway for *Cent* magazine. We can see that even heavyweight fabric billows away from the body in underwater conditions. Here, the feminine silhouette is preserved by a button at the waist. © Zena Holloway (www.zenaholloway.com)

underwater environment to expand the silhouette outward, and hence many such images feature unusually long hemlines to increase the quantity of floating fabric and enhance the spectacle of the underwater scene. A long skirt will billow outward, away from the body, just as it would in microgravity, creating a soft halo around the body for ethereal effect. The effect is to extend the presence of the body.[5] Underwater, and in microgravity, length is not a linear, vertical measurement as it is in Earth fashion, and nor does it equate to coverage. Indeed, a long, loose garment may be more exposing than short, tight clothing. An empire line gown, for example, might be lifted above the waist, as the loose fabric floats outward. The effect is illustrated in a series of underwater photographs by Michael David Adams, *Fire in the Sky*, in which floating bodies appear to be encircled by fabric rather than dressed or draped in it (see Figure 2.6).

The effects of hydrodynamic drag are visible in underwater photography whenever the subject, or the surrounding water, is in motion. When an object, including a dressed body, moves through the water, water is displaced around

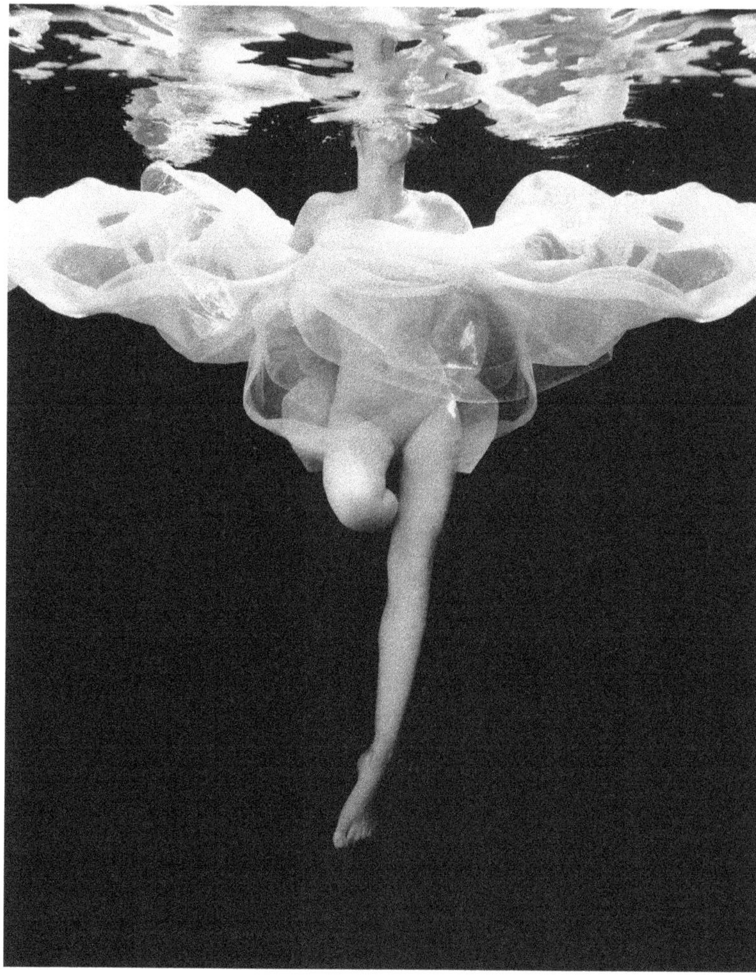

Figure 2.6 Photographer Michael David Adams's underwater photographs. He captures how fabric tends to drift away from the body, spreading outward to create a wide circumference around the body. Here, the body seems surrounded by fabric rather than draped in it. Rather than hanging downward, a long skirt will billow outward underwater, away from the body, just as it would in microgravity. The effect is to extend the presence of the body. © Michael David Adams www.MichaelDavidAdams-Underwater.com

it, creating a flow around the body, and turbulence behind it. Since it is difficult to hold the body still underwater, underwater fashion displays often present the body and garment in motion, and so prompt consideration of both static and dynamic drape. It is immediately visible in surveys of underwater fashion shoots that garments behave differently depending on how their wearers move. Since turbulence is created behind the moving body, it may affect the hem of the skirt when the model is swimming forward, or the shoulders when the model is swimming backward. Drag can force the fabric of a garment toward the body, so that the garment appears to become tighter, or it can force the fabric away from the body, depending on the direction of the flow. The more a model moves, the more the garment is affected by drag. If a body is propelled forward, as when a clothed model dives into the water, drag creates a flow that starts at her head and travels in the direction of her feet. This flow causes the fabric of her dress to cling to her body. At the hem, which has been dragged past her feet, turbulence causes the fabric to bunch and swirl. This turbulence is caused when flows that have traveled around her body meet in the water just beyond her feet. Additional turbulence can also be created by the fabric of the garment. A highly textured surface or wrinkled surface is less streamlined, and so creates additional turbulence (Watkins and Dunne, 2015, pp. 336–337). The effect of drag makes the underwater environment a less authentic simulation of space when objects are in motion. In order to make best use of water as an environment for testing spacewear, efforts should be made to minimize movement.

An underwater catwalk show for Russian designer Jana Nedzvetskaya illustrates the effects of hydrodynamic drag, as well as the contrasting forms that a garment can adopt when worn in air or water. Nedzvetskaya debuted her *Harmagedon Psalm 36:29* collection (Spring/Summer 2015) in a show that combines a traditional catwalk show with synchronized swimming and an underwater fashion display. Models wearing floor-length silk chiffon gowns stride to the end of Nedzvetskaya's runway, then plunge downward into a swimming pool. The show allows the garments to be seen in two different environments. In the air, they drape downward, and underwater they billow and swirl around the models' bodies. Unlike the predictable, oscillating motion of the skirt that is created as the model walks down the catwalk, the models' underwater acrobatics causes the silk to flow and twist in several directions, generating a diverse range of silhouettes. The various dresses show the extent to which length can be transformed into circumference, at the expense of the wearer's modesty. The floor-length empire line wedding dress that concludes Nedzvetskaya's show is lifted as she enters the water, so that her legs are entirely exposed.

It is possible to develop an appreciation of the different effects of water and microgravity by observing YouTube footage of astronaut Frank De Winne handling empty clothes as he explains how clothes are reused rather than laundered in space.[6] In contrast to the footage of Nedzvetskaya's underwater fashion show, in which garments are shown sweeping through the water, footage of De Winne handling clothes on board the ISS reveals that in microgravity a floating garment remains surprisingly static and stiff. Similar behavior is seen in footage of British astronaut Helen Sharman, who was recorded wearing a frilly, pink blouse that she had received as a gift during her mission to the ISS. The recording shows the sleeves billow outward from Sharman's arms, so that they resemble leg-of-mutton sleeves popularized in the 1820s, and retain this form as she moves through the cabin. Worn by a model moving through an underwater environment, such sleeves would be flattened against the direction of travel, or swirled by turbulence. Were that model to stop swimming and remain still, the sleeves would come to rest slowly and adopt a more uniform shape. It is in these moments of stillness that an underwater garment most closely replicates the behavior of clothes of astronauts on board the ISS. Water, then, may be used to gain an inexact impression of what a garment may look like in space, and when real weightlessness is not achievable it offers an inexpensive alternative.

Free-falling fashion models

This footage of Helen Sharman showcasing her blouse to crewmates raises further questions about how a garment might be encountered when its wearer is not fixed at the same orientation as the viewer. This in turn invites us to consider what form a fashion show might take in space. In microgravity, "the laws of bodily orientation . . . are suspended" (D'Aloia, 2012, p. 222). There is no up or down, "no ground on which to fall and finding a moment of stability relies of a precarious conjunction of balance and movements" (Wood, 2014, p. 443). The video shows Sharman propelling herself through the cabin with her hands. Rather than parading past the audience as would a model on a catwalk, revealing the garment from front to back at the model parades and then turns, Sharman glides above the camera so that her outfit is revealed gradually from shoulder to toe as she passes. The garment is encountered shoulder-first, as Sharman propels herself head-first toward the camera before it pulls away and she glides past.

Those who experience weightlessness need to become accustomed to a new "spatio-temporal logic." Videos recorded on board the ISS show astronauts and cosmonauts struggling to maintain their orientation, upright relative to the camera, for the duration of their live broadcasts to Mission Control. Similar scenes are seen in photographs of passengers experiencing weightlessness on board a zero-g flight. Even when efforts are made to remain upright relative to the airplane's floor, bodies uncontrollably drift and revolve. Space tourist Charles Simonyi recalls his experiences as two-time commercial visitor to the ISS, having to constantly reorient himself. Navigating through the cabin, he found that "your wall becomes your floor" [*sic*], and what may be one astronaut's floor is another astronaut's ceiling. Richard Garriott, whose privately funded trip to the ISS took place between Simonyi's visits, described eating dinner on board the ISS with six crew members, all making efforts to fit around a small dining table. "The dinner table is usually full with four or five people right-side up. Then one or two people the other way, using the ceiling as the floor," he recounts (Fisher, 2009, p. 66).

The lack of a clearly identified floor, and the inability of astronauts to ground themselves on a surface, leads to changes in behavior that need to be considered by spacewear designers. In her study of the depiction of weightlessness in science-fiction film, Adriano D'Aloia (2012, pp. 224–232) observes that astronauts ground themselves by making contact with any solid surface, and that surface is very rarely beneath their feet. NASA astronaut Jerry Ross observes that it is frequently the hands, rather than the feet, that make contact, thus relocating the ground to beside, not below, the astronaut (Herbert, 1999). The body's orientation, relative to its surroundings, varies frequently, as does its orientation in relation to other objects within the same weightless environment. Passengers my approach each other from above or below, thereby accessing views of the body or garment that are not ordinarily seen on Earth. When planning for future fashion shows for space, spacewear designers may need to consider the impossibility of a catwalk parade in a weightless environment. The audience would not need to sit on the same plane as the runway. Indeed, the audience could be above, below, or circling all the way around the runway, so that their own bodies are oriented differently to the bodies of the models. Without a floor, the models would need some way to propel themselves through the cabin, with locomotion that originates from their hands rather than their feet (see Chapter 4).

When spaces on Earth are designed with unusual orientation, viewers do not tend to experience the same sense of ambiguity or disorientation that they might feel when weightless. Even when an object or scene is inverted, it is possible to recognize that it is the object, not the viewer that is upside down, as a natural sense of balance communicates as sense of uprightness, and his or her own feet are, in most cases, planted on the ground. Nonetheless, inverted scenes and objects do create visual experiences similar to those that we might experience in weightlessness, granting access to unfamiliar views and perspectives. In 2005, Viktor & Rolf opened an upside-down boutique in Milan, in which the neoclassical interior has been inverted, so that the floor resembles the ceiling and vice versa. In the window, a rigid chandelier appears to hang upward from the floor, and above it, a chair stands seat-down on the ceiling. In the interior, the ceiling is laid with parquet floor tiles, and archways extend downward. On an upside-down plinth, a vase of flowers has been stiffened so that petals and leaves appear weighted against the force of gravity. Within this uncanny scene, clothes hang the right way up, evidencing the artificiality of their surroundings. The apparently contradictory visual cues provided by the assortment of upside-down and right-side-up objects provides some insight into what it might be like to enter a boutique on board a future commercial space station.

Unlike the rigid interior of Viktor & Rolf's boutique, their *Upside Down* collection drapes differently when inverted. In Earth's gravity, there is a visible difference between garments that are upright and upside down. Unless it is stiffened, fabric is pulled downward by gravity. It drapes to form a changeable silhouette that can be malformed by gravitational forces if its orientation shifts. The *Upside Down* collection was displayed "first one way and then the other" at Viktor & Rolf's

Spring/Summer 2006 show (see Figures 2.7 and 2.8). All of the gowns were designed so that they could be worn "bottom up" or "bottom down," with hems pinned together at the shoulders (Evans and Frankel, 2008, p. 172). Worn the right way up, most of the gowns narrow at the waist, and fall against the body in an hourglass silhouette. The same gowns, worn upside down late in the show, appear to gain volume at the torso and shoulders where gravity causes the fabric to flare outward.

The absence of a sense of "upright" in microgravity renders many fashion industry practices ineffective. A tendency to describe, design, and display garments according to the form that they take in an upright position makes the fashion industry unprepared for the disorientation that space tourists will encounter. Garments are manufactured from drawings that assume the wearer will remain in an upright position and, when those manufactured garments are displayed to consumers, they are typically displayed on upright models and mannequins. There is, in design, manufacture, and display, such a tendency toward the upright view that features viewable from above or below are neglected.

Window displays have exposed consumers to alternative views by positioning mannequins as if flying or floating weightlessly. In Alber Elbaz's 2015 Icarus-themed window display for Lanvin, Paris, mannequins are suspended as if falling (Fleischer, 2015). Some are seen moments from impact, shielding their bodies from the anticipated impact. Others are dancing through the air. All are oriented in ways that expose parts of clothes that are not normally seen: the soles of shoes, the lining of skirts, the inner thigh seam. Gravity pulls the fabric of their clothes in directions that distort the silhouette: skirts drape perpendicular to the line of the mannequin's body; blouses bulge at the shoulder. The soft fabric drapes, forming new silhouettes that do not match the shape that these garments would form had these mannequins been upright. These unexpected shapes draw attention to the qualities of the fabric, highlighting its capacity to drape in soft curves around the body.

Fashion brands have also found ways to exhibit their garments on unconventionally oriented catwalks. Jochen Schweizer has arranged vertical catwalk shows for Espirit (Munich, 2002), Marshall Field (2003, see Figure 2.9), and Target (New York, 2005) and has inspired more recent vertical runway events by other brands including Gucci to Tokyo (in collaboration with Shogakukan, Tokyo 2014). This German agency unfurls cloth runways from the roof of multistorey buildings, creating a vertical catwalk that extends down the façade of the building. They then provide acrobats to act as models, belaying from the roof, propelling themselves away from the building whenever their feet make contact with the surface, so that it appears they are striding, "face down," down the catwalk's vertical surface (Ganz München, 2002). The lack of gravitational force on their shoulders allows the harnessed models to propel themselves

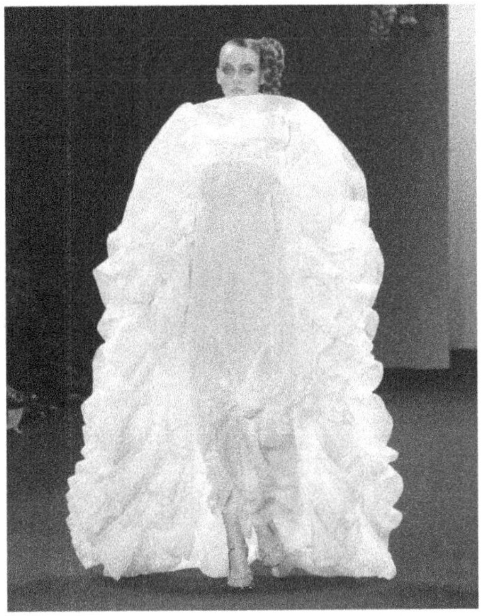

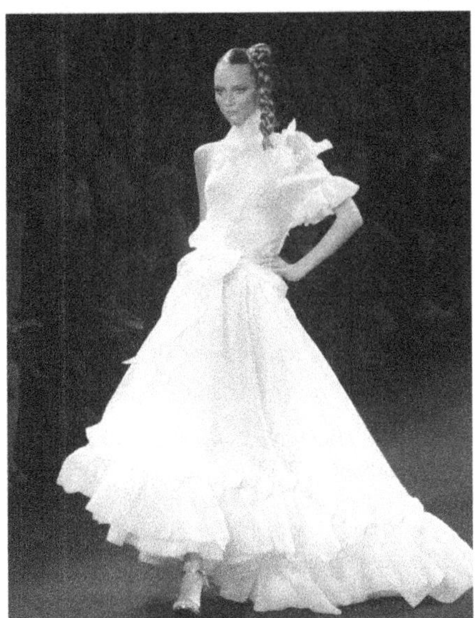

Figures 2.7 and 2.8 In Viktor and Rolf's Spring/Summer 2006 collection, Upside Down, the same gown is worn upside down (top) and the right way up (bottom). © Jean Baptiste Lacroix/FilmMagic

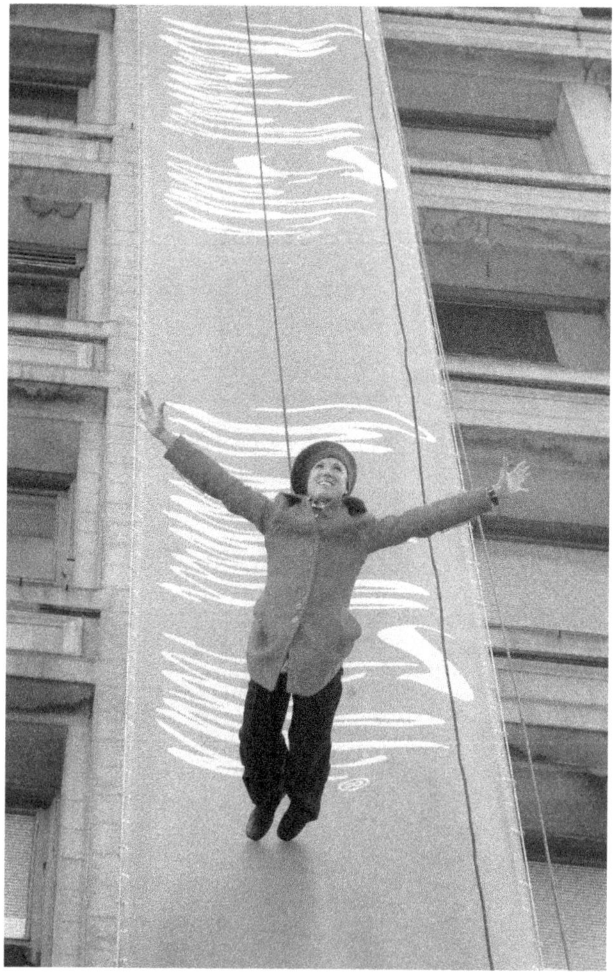

Figure 2.9 A fashion model works her way down a vertical runway constructed on the side of the Marshal Fields store in Chicago, 2003. Models in the vertical fashion show stepped off the eleventh floor of the building with the aid of rigging and rappelling gear. As gravity acts downward, it forces the model's coat into unfamiliar shapes. Here, additional volume is visible just below the collar, and at the waist the hem falls downward, perpendicular to the model's horizontal body. From this angle, the extra bulk created at the shoulders and collar appear very similar to that which can be seen in images of astronauts on board the ISS (see Chapter 4). © Scott Olson/Getty Images

outward as they descend, taking giant, slow-motion leaps that are not dissimilar to those familiar from footage of the moon landing.

Worn in these horizontal positions, a garment is reformed. Gravity exerts force perpendicular to the orientation of the garment, so that it drapes away from the body. For Shogakukan and Gucci in 2014, models descended and then ascended a 42 meter-high vertical runway in Tokyo. Wearing the same garments while descending and ascending, the models' bodies were oriented with and against gravity, so that their garments draped differently on each pass. In a vertical catwalk show for New Face in Weimar, in 2013, models began the show in a handstand pose on the roof, and were then lowered over the edge until they were hanging so that their bodies were parallel to the wall of the building. From there, they flipped over, propelling themselves horizontally into the air as if jumping off the ground, and then began their descent. These acrobatics result in the clothes being displayed from every angle, and draping over the body in a variety of ways. As the models' bodies revolve, the force of gravity acts on their garments so that the drape continuously undulates around the body.

Many common approaches to the form of a garment are incompatible with any orientation other than vertical and upright. The garment that is designed to be viewed on an upright body may look very different when viewed from above or below, or any of the angles in between. Clothes are often defined by their silhouette, that is, the vertical contours of the garment that give it a distinct ratio of bust to waist to hip to hem. Silhouette, more so than other features of womenswear, has been used to represent the evolution of fashion, with developments defined in terms of silhouette, as was the case with the empire line of Regency-era gowns, the drop-waist of Modernism, or the A-line of Dior's New Look. Early fashion illustrations frequently depict front and back views, flattened onto the page. Various developments in the history of fashion have given illustrators cause to depict the side view, as was the case in *Harper's Bazaar* illustrations of the latter half of the nineteenth century, when the shape of bustles could not be clearly depicted in front or rear views (see, for example, the cover of *Harper's Bazaar*, October 24, 1885), or at the turn of the twentieth century, during the introduction of the s-bend corset (see, for example, the cover of April 7, 1900). The term "the new silhouette" has been used to introduce developments in fashion for more than a century. *Vogue* used the phrase in 1905 (April, p. vi) to describe the latest gowns from Pairs, with their clearly defined waists, and then again in 1913 (April, p. 44) to herald the emergence of the modernist, tubular silhouette. As the shape of womenswear continued to evolve throughout the twentieth century, each new era was welcomed with a celebration of another "new silhouette," from the New Look (September 1947, p. 141) to the "slim skirts, narrow pants, and asymmetric gowns" of Fall 1996 (September, p. 578). Sue Jenkin Jones (2011, p. 280) proposes that the importance of the silhouette derives from its immediate impact, "seen from a distance and before the details can be discerned."

While the silhouette is no doubt an important feature of a garment, descriptions of silhouettes assume that the garment is encountered in front, side, or rear elevation. However, in a microgravity environment, as in the footage of Helen Sharman showcasing her frilly blouse on board the ISS, the elevation view is rarely seen. Silhouette changes along with the wearer's orientation, as volume is concealed and revealed as the dressed body is rotated relative to the viewer. This preoccupation with the silhouette, as it is used to represent the side or front elevation view, has led designers to neglect the aspects of design that are visible from other angles, such as when a garment is viewed from above or below. Little or no consideration is given to the alternative silhouettes that could be encountered when a garment is viewed from above or below, as is likely in weightlessness. It is no coincidence that when fashion photography started to feature more dynamic angles, these experiments coincided with the Space Age.

The Space Age and developments that followed disrupted the primacy of the elevation view. "Verticality" has historically dominated human understanding of the world around us, dictating how we order physical spaces, objects, and ideas. "Vertical sovereignty splits space into stacked horizontal layers, separating not only airspace from ground, but also splitting ground from underground, and airspace into various layers" (Steyerl, 2011). The advent of aerial photography and, later, satellite imagery, introduced new ways of looking at the Earth. Writing on the increasing ubiquity of aerial views in satellite images, such as those we now see in Google maps, Hito Steyerl (2011) concludes that "we are growing increasingly accustomed to what used to be called a God's-eye view." The "stable and single point of view" that is depicted in linear perspective or elevation plans "is being supplemented (and often replaced) by multiple perspectives." Steyerl examines the importance of the horizon as an orienting cue, essential in navigation, but absent when we journey underwater or into space. Aircraft expanded our methods of orienting ourselves and other objects. Satellite images have prompted "a radicalization . . . of the paradigm of linear perspective" and, in the loss of a horizon, the "departure of a stable paradigm of orientation." This is not merely a visual phenomenon. Astronauts describe a physical "sensation of tumbling upside down" in microgravity (Clark et al., 2015), so that orientation perspective is disrupted even when artificial orientation cues are introduced. The Space Age may therefore be seen as an age of disorientation in which the stable, elevation view is exposed as just one of many possible alternatives.

In Peter Knapp's photographs of Andre Courrèges's 1965 collection for *Elle*, models are depicted as if weightless, floating across the spread and off the edge of the page (Figure 2.10). The camera is positioned beneath, so that the photographs capture the soles of the models' shoes and the lining of their skirts. Each model floats at a different orientation and distance, as if uncontrollably drifting around the studio. Some are cropped by the edge of the image, much like the subject of a Toulouse Lautrec poster, as if they are unintentionally floating out of the frame. In one image, a male model's jacket is held open, as if it is in the process of floating away from his body. The backdrop is unlit, resembling the darkness of space. These photographs do not show the silhouette of Courrèges's garments, at least according to the common use of the term. Photographed

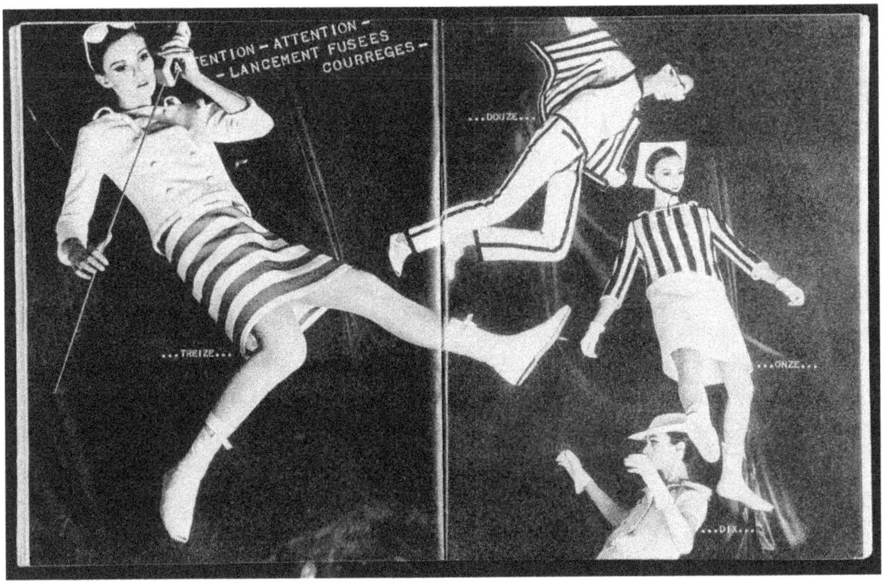

Figure 2.10 Peter Knapp, *Courrèges for Elle*, 1965. Knapp's photographs depict models apparently floating through space, so that they are viewed at a variety of different angles. From these angles, cuffs and hems appear as ellipses, the lining of jackets and skirts is visible, and the soles of shoes are on display. The image provides some insight into the range of different viewpoints that fashion designers will need to consider for the weightless environment. © Peter Knapp/Audrey Hoareau

from below, the models' hats appear as circles and squares, reduced to the shape of their brims, and the rings of their cuffs encircle their hands. Boots create focal points, illuminated in the foreground as the models' upper bodies recede. With legs flayed in all directions, we are shown the boots from every angle, particularly from underneath their flat soles. These angles expose shapes and contours that may not form part of a designer's original illustrations, but that are taken for granted in fitting a garment to a body. The shapes that allow a body to pass through a garment—the loops of the cuffs and the hem—dominate.

In Knapp's image, as in space, the absence of a visible horizon is disorienting. There is no visible divide between the upper and lower parts of the image. The only lines present are those printed on the garments, which extend in different directions across or along the models' bodies, and, due to the haphazard arrangement of those bodies, those lines are neither parallel to the top nor to the sides of the image frame. The high contrast of white fabric against a black backdrop makes the hems clearly visible, but these appear as ellipses, not as the straight, horizontal lines that one might expect in an illustration of the same garments. When depicting the silhouette of a garment, one might consider the hems equivalent to horizons, providing visual information that orients the garment. When we observe a flat horizon, we understand that it is a translation (visually flattened) of the curve of the globe. Likewise, we understand that the hem that is depicted in silhouette as a straight line is representative of a shape that curves around the body. If we are relocated to space, where the horizon is no longer available as an orienting aid, we become more aware of alternative perspectives, such as the God's-eye view captured by satellites and viewed through observation windows on board the ISS. In this environment we also encounter the worn garment in alternative orientations. The Earth's horizon becomes a circular segment, and a hem, as in Knapp's photograph, becomes a complete ellipse.

Such is the primacy of the elevation view that it continues to inform the ways in which garments are constructed, on the same planes as these elevation-view illustrations. Seams conventionally run horizontally or vertically. Even asymmetric hems tend to meet at straight, vertical seams. Pattern pieces divide the body into a vertical grid. After manufacture, clothes are, for the most part, designed to lie flat on their front or back, flattening like an envelope, or to hang vertically for storage. At every stage of design, manufacture, and display, there is reference to the flattened silhouette. There are, of course, exceptions. Issey Miyake's *132 5* are designed to be viewed from above when unworn, folding flat into a variety of origami-inspired shapes. Miyake's designs flatten more like an accordion than an envelope.

Designer Julian Roberts acknowledges the significant role that gravity plays in dressmaking, and in particular the way in which the force of gravity is responsible for transforming a flat plane into a three-dimensional garment. Roberts's "subtraction cutting" is a dressmaking technique that involves placing the patterns for front and back bodice apart on the same plane of a flat tube of fabric, then removing the space between them before pulling the two separate bodices together, thereby drawing the remainder of the tube into an irregular skirt (Figure 2.11). Roberts (2013, p. 22) contrasts the "elevated view" taken by most fashion designers with the "aerial view" and "plan view" that are the starting points for the construction of his garments. He approaches the rest of the dressmaking process much like a topologist might consider the elastic space between two points, by cutting holes at various points on the tube and then stitching them together. These holes eventually form the tunnel through which the wearer's body passes.

Roberts (2013, p. 27) presents the example of a skirt, which, when presented from an elevation view, may resemble a rectangle or trapezoid, but when viewed from an aerial view resembles a flattened donut—a large circle with a smaller circular hole in the center (see Figure 2.12). He envisions that the pattern for a skirt could be cut according to either of these two views: conventionally, as two trapezoid shapes, front and back, stitched together down the sides to form vertical seams; or as a large circle with a smaller circle subtracted from the

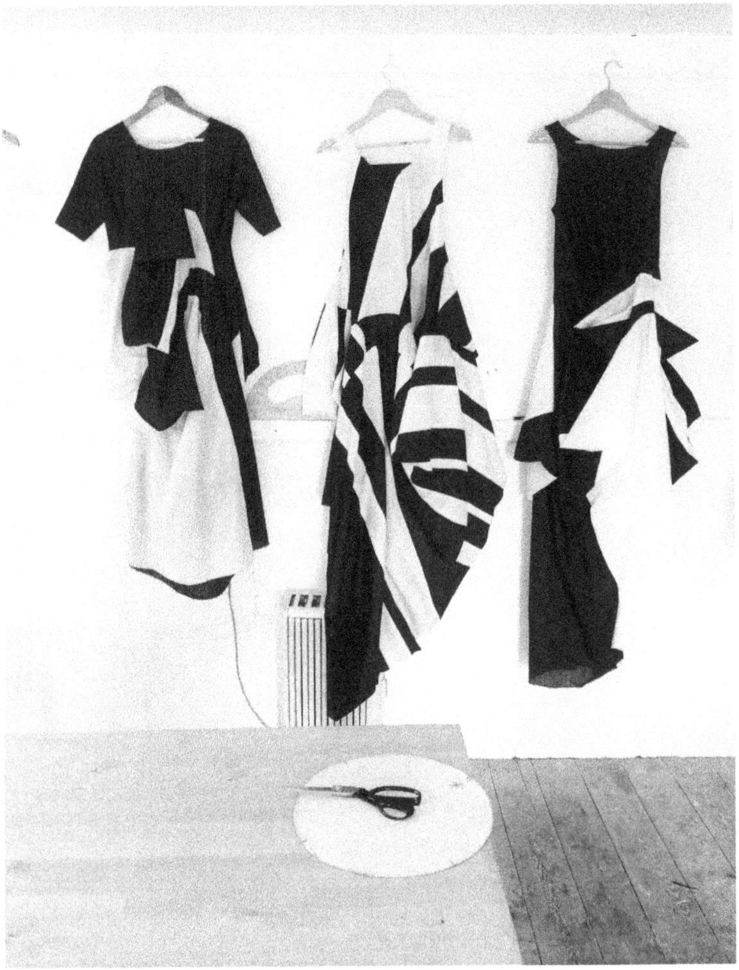

Figure 2.11 Julian Roberts's dresses made using the "subtraction cutting" method. The irregular drape is caused by the unaligned placement of bodice pattern pieces. © Julian Robert's design studio, Brighton 2014.

middle, with no seams. This second pattern, resembling the aerial view, is pulled into a three-dimensional skirt when the force of gravity works to pull the outer edges downward while the central hole is held aloft by the wearer's body (p. 28). Roberts also describes patterns in which the aerial and elevation view are mixed together, in a plan view (p. 29).

Familiar shapes become unrecognizable when viewed at unfamiliar orientations. Roberts's aerial view reveals the skirt as two concentric circles. In practice, as can be seen in Knapp's Space Age photographs, these shapes tend not to be perfect circles (unless worn with a crinoline). The fluidity of fabric

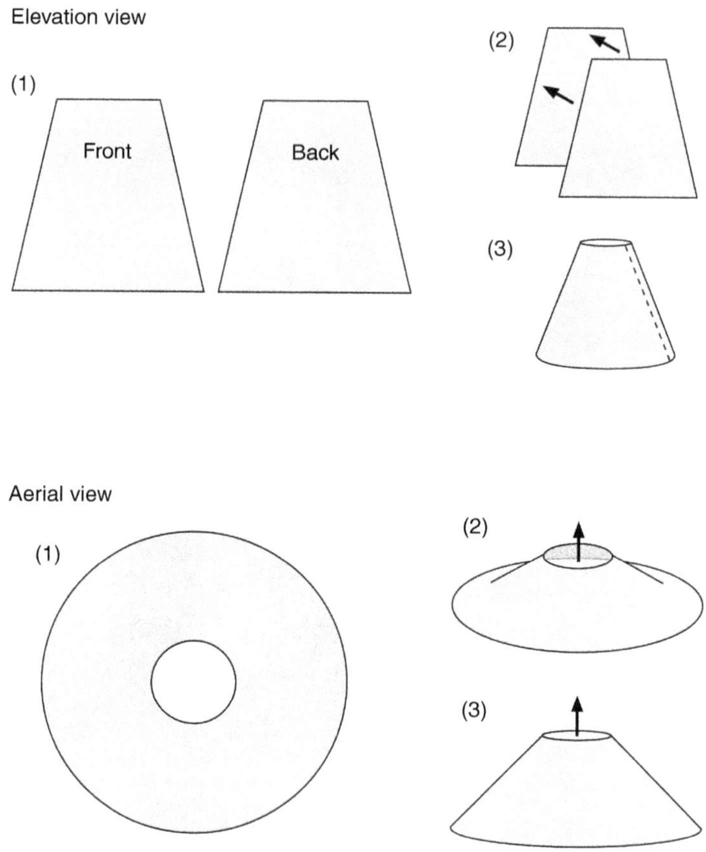

Figure 2.12 Julian Roberts's alternative approach to making a skirt. The conventional, elevated view (Figure 2.11) involves two pattern pieces, one at the front and the other at the back, stitched together at the sides to form vertical seams. His alternative aerial view (Figure 2.12) shows how a skirt can be created from a single pattern piece with an area subtracted. The flat donut shape is transformed into a skirt by the force of gravity, pulling downward at the hem. Adapted from Roberts (2013, p. 28).

transforms the outer circle that is the skirt's hem into a distorted loop, which undulates as the wearer moves her legs. From the same orientation, a pair of trousers might present itself as a pair of smaller circles (the hems at each ankle) enclosed by a larger circle (the waist). From below, a shirt might present itself as a pair of small circles (the cuffs) on either side of a larger circle (the waist). From each of these circles, a part of the wearer's body extends. In order to design a garment that is to be viewed from above or below, one must be aware of these shapes, and consider them as significant as the front or side silhouette.

Prompted to consider the aerial view during the design process, spacewear designers may begin to create designs with more varied, and more interesting, aerial silhouettes. In many existing garments, little would be revealed by displaying an aerial silhouette. Silhouette recognition relies on the ability to imagine a flat shape transformed into a three-dimensional object. A certain amount of detail is required for the mind to succeed in this translation from 2D to 3D. Since the aerial view of a garment tends to be relatively featureless, it is inadequate to communicate the form of a garment without the provision of some additional information. The silhouette of a body in a gown, for example, drawn as if viewed from below, may present itself as a featureless circle, indistinguishable from the silhouette of a sphere. Such featureless aerial silhouettes may be a consequence of designers' overreliance on the fashion silhouette and the elevation view. While it may be possible to infer the shape of a garment in aerial view by visual information provided in the elevation view, it would be wrong to discount the value of considering the aerial view during the spacewear design process. Spacewear designers will face the challenge of making the aerial silhouette as interesting as the front or side silhouette. By establishing the importance of the aerial view, new design possibilities are revealed.

However vital it may be to find an effective method of depicting the form of a garment when viewed from above or below, it is equally essential to acknowledge that aerial and elevation are not the only views. Arthur Woods (1993, p. 299), creator of *The Cosmic Dancer* (see Chapter 1) notes that "a sculpture floating in a zero-G environment could be viewed from an infinite number of perspectives or angles." The same is true of all three-dimensional design. When designing for weightlessness, we must consider that the body may be viewed from all orientations. Establishing the importance of the aerial view does not eliminate the need to also consider the elevation view. The weightless body exposes a garment to viewers from all orientations, and so all of those views must be consciously considered by the designer. We must design from bottom to top, and vice versa, as well as from front to back, and all other angles in between.

We must also consider the relationship between connected contours that may be visible at these different orientations. They may be drawn in two separate images, but the elevation view and aerial view are not disconnected. Though only one of these views may be visible at any one time, the shape of one is often significantly related to the other. For example, a wider hem depicted in silhouette or elevation view would result in a larger circumference of circle in the aerial view. Moreover, it is unlikely that a garment will ever be viewed directly from above or below, or front or back. The binocular disparity that allows us to see in three dimensions, combined with the body's tendency to adopt a neutral position in microgravity (see Chapter 4), with the body slightly bent, means that multiple aspects may be visible simultaneously.

A method of three-dimensional depiction therefore needs to be developed that allows for contours to be designed at any orientation, and that intelligently adapts the interdependent contours that might be viewed from alternative orientations. The value of designing for multiple orientations may prompt designers to consider the potential of previously unseen shapes and surfaces. At its most basic, consideration of the view of a garment from below may prompt designers to more carefully select a lining. More innovatively, designers may begin to consider adding new contours for display from a variety of orientations.

3
COMMERCIAL SPACEWEAR

The increasing commercial interest in space has led to a shift in approaches to spacesuit design. Safety and functionality were the primary concerns of spacesuit engineers throughout the twentieth century. In the new, commercial space age, when popular opinion and public investment are key, competition between designers and manufacturers has led to an increased concern for aesthetics. Commercial spaceflight operators have approached the design of spacesuits and flight suits as an integral part of their activities in the lead-up to launching tourists into space, and as a way of reinforcing their brand identities in a fiercely competitive market. Designing suits for both their crew and their passengers, commercial operators evidently consider a branded flight suit an integral part of the spaceflight package, with Elon Musk fully aware that SpaceX customers' choice of spaceflight provider may be driven in part by a desire to look "stylish" (Cuthbertson, 2016).

In the era of mass space tourism, it is likely that there will be a convergence of Earth fashion design and the established practices of spacesuit engineers. Commercial spacesuit engineers have already begun to operate like fashion designers, taking inspiration from fashion houses by producing ready-to-wear lines as well as bespoke "couture" items. Commercial spaceflight companies such as Virgin Galactic, SpaceX, and Spaceport Sweden have sought to employ fashion designers and costume designers to develop suits for their future passengers and crew. Meanwhile, boutique engineers, such as Brooklyn-based Final Frontier Designs, have anticipated a need to offer bespoke or off-the-shelf spacewear and spacesuits to cater to individual clients and consumers. All of these designers have grappled with the complex challenge of balancing functionality and aesthetics.

There are two main categories of worn artifacts for space: spacesuits, engineered to protect the body from the dangers of space; and other, everyday clothing, worn on board a spacecraft. The concerns of spacesuit engineers differ significantly from those of the designers and buyers who create or select clothing for intravehicular activity (see Chapter 4). While spacesuit engineers must protect the body from radiation, pressure, and extremes of temperature, the cabin of a spacecraft protects passengers from these dangers, enabling the wearing of everyday, off-the-shelf clothing. However, as spacesuit engineers begin to develop greater concern for aesthetics, and fashion designers begin to incorporate wearable technology, there are signs that the interests of these two fields will align. There are commercial interests in clothing worn for both intravehicular activity (IVA) and extravehicular activity (EVA), with specialist engineers working to perfect the technologies that protect the wearer from the vacuum of space, and fashion designers working with commercial space organizations like Virgin Galactic to clothe travelers who will remain inside the spacecraft. Commercial spaceflight organizations have primarily focused their attention on the design of pressure suits and flight suits, to be worn inside a spacecraft. Pressure suits, which protect the body in case of unexpected depressurization of the vehicle, are a safety measure employed by existing government organizations for launch and reentry. It is not yet known whether commercial spaceflight operators will be required by regulating bodies to clothe their passengers in pressure suits (Webber, 2017, p. 144), leaving commercial operators to choose whether to focus their attention on pressure suits or unpressurized flight suits. To determine the needs of future space tourists, it may be useful for spacewear designers to consider the full range of worn artifacts in use by NASA astronauts, which include spacesuits for EVA, and pressure suits, flight suits, and commercially available clothing for IVA. NASA's use of commercially available off-the-shelf clothing, such as polo shirts and shorts, illustrates the need for a spacewear market that extends to everyday wear for the weightless environment.

Form, function, and fantasy

Utilitarian concerns have driven the choice of clothing for space travel. Spacesuits have historically been developed by engineers, not designers, with the suit's primary purpose being to enable the survival of an astronaut during EVA. Aside from the preservation of the life, one of the most essential driving factors in spacesuit design has been the need for astronauts to carry out maintenance tasks and other labor while suited. The suit is equipped with pockets and appendages that enable those tasks. Spacesuits double as surfaces on which all manner of useful devices could be stored within easy reach. Tools are stuck, strapped, and tethered to the outside of the suit, ranging from cameras mounted to the chest to, in the case of Apollo 11 astronauts, checklists, stitched onto the wrist of the gloves (Hersch, 2009a, p. 14).

In order to access these tools and carry out their tasks, astronauts require some flexibility at the joints. Spacesuit engineers have sought to introduce flexibility without compromising safety, resulting in discomfort and restricted joint mobility. As Mark Timmins (2010, p. 198) observes, spacesuits are "machines not for living but for accommodating and only just tolerating humans." Spacesuits are unwieldy objects. Hard Upper Torsos (HUTs) have a rigid shell, and even Soft Upper Torsos (SUTs), which are soft when donned, become stiff when pressurized (Ferl et al., 2006, p. 1). Soft arms and legs, attached to the upper torso become virtually rigid when pressurized, requiring pleated joints and rotational bearings that allow for mobility (Schmitt, 2001, pp. 23–24). One astronaut likens the experience of the suit becoming pressurized to operating in wet concrete, which hardens as the suit is pressurized (Gast and Moore, 2011, p. 320). The bulky nature of spacesuits, a consequence of the need to cater to an extraordinarily dangerous environment, have necessitated a focus on function over form.

Modernist, utilitarian principles were at the heart of twentieth-century space design. Adams and Jones (2014, p. 72) observe that the dominant design influence on space architecture has been modernism and, in particular, modularity. When developing plans for the International Space Station (ISS), "NASA's best thinkers . . . came to the conclusion that a diligently modular approach to design and execution would be the best guarantee of success for America's Space Station" (p. 73). Modularity would allow for the completion and expansion of the ISS over the following decades, accommodating "unknown or uncertain components" that had not yet been designed at the time of the ISS's original launch. In recent years, its modular design has enabled commercial space agencies to consider extensions that serve commercial purposes, including Bigelow Aerospace's inflatable structure, BA330, intended for use as a "space hotel" (Doule, 2014b, p. 69). Modularity has been at the core of spacesuit design too, with suits typically made from interconnecting parts. Waist-entry suits, such as the Shuttle EMU, are split into upper and lower halves, connecting at the waist by means of a Body Seal Closure.[1]

Modularity gives the EMU several practical advantages over a single unit suit. First, it increases flexibility of use, with the possibility of exchanging component parts to suit new users or mission requirements. Second, it enables donning and doffing in microgravity, as act that would otherwise be much more difficult. The means by which the separate parts connect dictates, and is dictated by, the shape and rigidity of the spacesuits various parts, and this is in turn dictated by the method by which the astronaut dons the suit. Depending on the style, space suits are donned and doffed through waist entry, rear entry, or zipper closure (Ferl et al., 2006, p. 1). These two parts are pulled on, one at a time, onto the top and bottom half of the body. The wide circumference of the waist is necessarily dictated by the need for donning and doffing. Rear entry suits, such as the Russian Orlan suit, typically have a hinged door beneath the backpack. These suits have more mass in the area above and behind the shoulders, as the astronaut's entire body must pass through this opening. The "ability to don the spacesuit quickly and safely is critical to maximizing overall productive EVA time," but since the ease of donning is related to the size of the opening, bulk and maneuverability must be compromised in order to improve the ease at which a spacesuit is donned or doffed (Graziosi, Ferl, and Splawn, 2005, pp. 1–5).

The process of donning a spacesuit bears little resemblance to the everyday dressing actions that are performed on Earth and, as any with many other actions

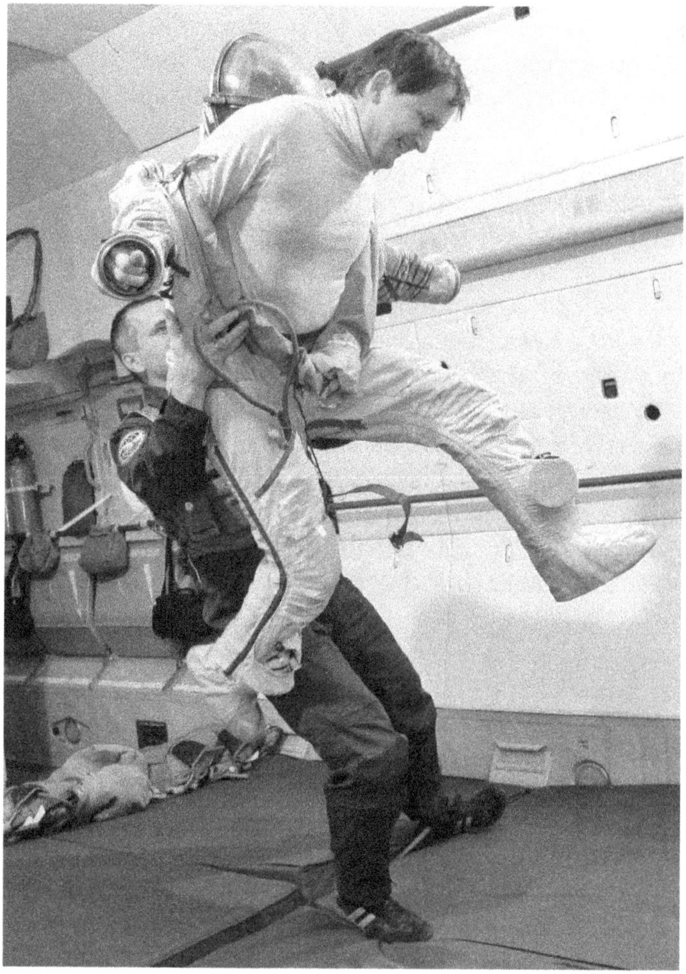

Figure 3.1 A Russian instructor (left) assists former Microsoft software developer and space tourist Charles Simonyi to don a space suit during a parabolic flight aboard a zero-gravity simulator, before his journey to the ISS in April 2007. The image illustrates the difficulty of dressing when gravity pulls neither the body nor the garment toward the ground. The problem of dressing and undressing in weightless environments is compounded by the bulkiness and stiffness of a spacesuit, which is difficult to don even within Earth's gravity. © MAXIM MARMUR/AFP/Getty Images

performed in microgravity, requires training specific to the type of spacesuit that is to be worn (see Figure 3.1). Dressing and undressing on Earth are characterized by vertical movements, as parts of a garment are pulled against gravity, or released and allowed to fall onto the body or floor. It is common for Earthwear to have entry and exit points at the top and bottom (collars at the top, hems at the bottom, through which the body is threaded vertically), and for the gestures involved in dressing and undressing to be accompanied by stabilizing posture, with the body planted on the ground or seated. Rear-entry spacesuits respond to the possibility of entering a garment without the need to have one or both feet planted on the floor, and the weightlessness of the body overall. Rear entry, in which the folded body floats into the suit, head and feet forward with buttocks trailing behind, is, for astronauts, a solution to the problem of a bulky and inflexible waist closure. This is "not the most viewer-friendly activity," says astronaut Chris Hadfield (Savage, 2015) as he describes the undignified process of doffing a rear-entry suit that involves exiting the suit "butt first." For future spacewear designers, these alternative methods of climbing in and out of a suit illustrate the possibility of relocating entry points so that a garment does not necessarily need widened or elasticated openings at the collar, limbs, or waist. As in Earthwear, dressing and undressing can be made easier with modular designs, with gloves and boots that are attached separately. However, while modularity might ease dressing in some ways, an inability to control loose parts might also make it more difficult. Hadfield recalls how "putting on boots is the hardest thing in zero-G" because "they float all over" (Savage, 2015).

While it is important that future spacewear designs do not compromise safety, there are signs that aesthetic concerns are coming to the fore. There has been a shift from function to form in the design of spacesuits, prompted by increasing public engagement. As part of its drive to encourage audience involvement, in 2014 NASA invited audiences to vote on the design of their new generation of spacesuit, the Z-2. Illustrations of the proposed designs were posted on a NASA website, where the public were invited to cast their votes to select the suit that will eventually be worn by NASA astronauts in extravehicular exploration on the surface of Mars (see Figure 3.2). The functional aspects of the Z-2 suit were established in the design of the previous suit, the Z-1, which was named one of *Time* magazine's best inventions of 2012, thanks to its ground-breaking application of 3D printing in the formation of impact-resistance structures (NASA, 2014b). This design was lauded primarily due to its innovative functionality. The new design, the Z-2, incorporates similar technologies and, crucially for voters, it also looks stylish.

The decision to invite the public vote led NASA to incorporate elements of fashion, making it necessary for the suits to mirror aspects of contemporary fashion design, drawing on trends for vibrant colors and sportswear, combined with visions of the future in recent sci-fi film costumes. The website offered a

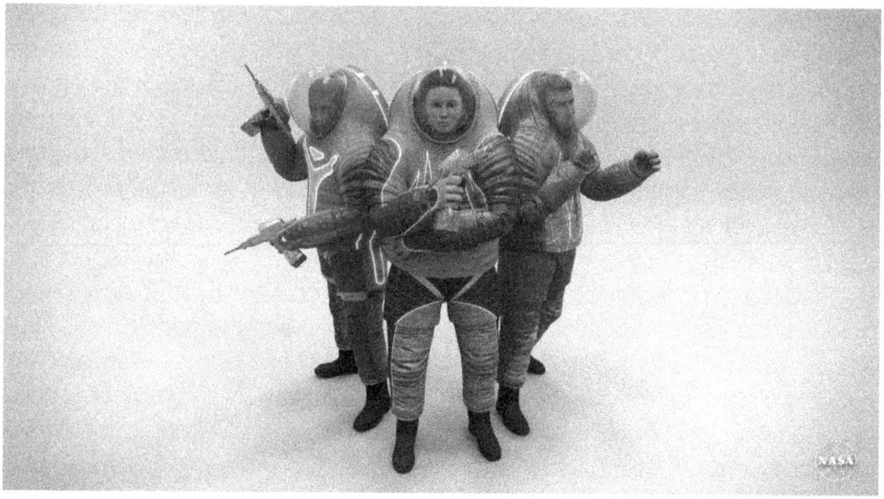

Figure 3.2 Three designs for NASA's Z-2 spacesuit that were offered for public vote. The winning suit (left), "pays homage to spacesuit achievements of the past while incorporating subtle elements of the future" (NASA, 2014d). © NASA

selection of three spacesuits, inspired by principles of biomimicry and trends in wearable technology (Holpuch, 2014). One of the designs, titled the *Biomimicry* suit, contains an electroluminescent wire, inspired by aquatic creatures, which decorates the suit in low lighting conditions. The *Trends in Society* suit—the most overtly aesthetic design of them all with its contrast stitching—is reflective of what everyday clothes may look like in the not too distant future, and takes inspiration from sportswear. The winning design, the *Technology* suit, incorporates an illuminated chest insignia.

Civilians voted for a Z-2 design that would be worn by astronauts, not a suit that they would wear themselves. Their concerns were, therefore, inevitably focused on the exterior features, and not the mobility, comfort, or general usability of the design. In terms of basic structure and functionality, the three designs offered to voters were identical, differentiated only by superficial aesthetic elements. The role of the public vote was therefore not to determine how the suit would be engineered, but how it would be decorated. The superficiality of these choices highlights both the limits of the audience's expertise (in contrast to the expert spacesuit engineer) and the importance of aesthetics in an increasingly commercialized field. It is reasonable to assume that most voters have no experience of space travel, and are far less qualified than NASA employees to make informed decisions about the functionality of any particular spacesuit. Therefore, in order to offer voters an ostensibly significant level of involvement, their influence must be restricted to superficial elements of style. Historically held notions about the nobility of sight have led to an overemphasis of the importance of the suits' visual features, making stylistic features appear more important than they actually are, and giving voters the impression that they are making a significant contribution to the future of space exploration. NASA's invitation to voters is of little significance to the astronauts who will wear the suits, but vital to the public perception that they are engaging with their audience. In an increasingly commercialized industry, NASA must stay ahead of the game not just in technical innovation but also in terms of public image.

This example highlights the vast difference between indirect public engagement with future space exploration, and direct public participation in future space travel. Commercial spacesuit and spacewear designers recognize the appeal of looking like an astronaut, even if that is the only accessible part of the space travel experience. An industry is emerging to cater to the needs of an audience for whom dressing up as an astronaut is a way to participate in space travel culture without taking part in spaceflight. Recognizing the appeal of activities even on the periphery of space travel, independent spacesuit designers Final Frontier Design (FFD) offer commercial customers a "Space Suit Experience." This two-hour session involves donning a space suit and participating in mobility and range-of-motion tests, in order to "experience what it's like inside a pressurized space suit garment." The experience permits a participatory fandom that is more advanced and exclusive than the activities on offer to the general public at science museums and space centers. Public support for FFD's products and activities illustrates a popular interest in commercial spaceflight from civilians who, currently unable to experience spaceflight themselves, seek to engage in the commercial space race indirectly. FFD (2017) describes its target market for the experience as "space enthusiasts" who cannot afford (or cannot wait) to purchase a ticket on board a commercial spaceflight (Kramer, 2014a).

The activity of dressing up in a spacesuit might be considered akin to cosplay, a "ritual of identification," which aligns the participant with an astronaut hero or heroine through procedural play (Norris and Bainbridge, 2009). FFD's customers might have engaged in the imaginative reconstruction of space narratives, such as playful reenactments of the moon landing during childhood, but FFD's "Space Suit Experience" is distanced from childhood play by its emphasis on authenticity. Customers don a fully functioning, NASA certified pressure suit of FFD's own design. Authenticity is offered not only in the spacesuit, but in the whole experience. Customers experience an "immersive event" that allows them to role-play as an astronaut (Messier, 2014). After body measurements are taken, as they are for real astronauts, participants wait while alterations are made to the suit. They then endure the claustrophobic experience of being inside a spacesuit as it pressurizes, first in standing position, and then in a launch simulator.

The company is given credibility by their contract to provide gloves to NASA, and their successful production of fully functioning flight suits for future space travelers. In these activities, too, they have been able to exploit public desire for indirect involvement in space travel. They have developed a small range of "garments for the future of space travel," including IVA suits for the commercial market, and is among a small number of spacesuit manufacturers catering "specifically for the needs of civilian space access," funded in part by public donations (FFD, 2017). FFD (2012) successfully sought crowdfunding for the production of their third-generation spacesuit, with 386 backers making individual notations ranging from $3 to $10,000 in value.[2] Each of these donations would be too small to enable participation in spacefaring activities, but they are collectively large enough to make a noteworthy contribution toward future commercial space travel. For a small cost, backers are able to count themselves as part of the space community, contributing to the growth of the commercial space sector. The success of the campaign evidences the continued public support for space exploration, for the good of mankind rather than solely for personal experience. Just as during the first Space Race, when public support led to a collective feeling that it was not just NASA but the human species collectively that had made a giant leap into the future, now civilians can support commercial space exploration financially as well as in spirit, through "citizen-driven" activities like FFD's campaign (Kaminski, 2016, p. 222).

FFD's spacesuits have been described as the "ultimate in bespoke couture" (Fernholz, 2015). They are no longer just uniforms for astronauts, but the ultimate wardrobe item for fashion-forward consumers. Their principal interest is in anticipating the needs of future space tourists (Gittleson, 2013) and, in this regard, they have identified a niche audience whose needs are not currently met by other manufacturers of wearables. Fernholz's (2015) comparison of FFD's spacesuits to couture clothing recognizes that similarities between the activities of small start-up spacesuit manufacturers and boutique clothing labels that cater to similarly niche audiences with aspirational designs. The high cost of manufacturing a spacesuit is prohibitive, in part because they must be custom fitted to the bodies of their small but exclusive clientele.[3] FFD subsidizes its operations by manufacturing cheaper replica items for display in museums and private collections. To that extent, their business model is not dissimilar to that of fashion brands whose couture collections "symbolically subsidize" and are "economically subsidized by" their ready-to-wear lines (Moeran, 2004, p. 275).

FFD, and competitors such as Orbital Outfitters, whose "space diving" suit is intended for high-altitude skydivers as well as "emergency backup" for space travelers (Weed, 2007, p. 55), represent a small part of a commercial spacesuit manufacturing industry that is dominated by more established manufacturers such as ILC Dover, David Clark, and Boeing. These companies are principally contracted by government agencies, and so the requirements of their spacesuits are determined by NASA. At present, government space organizations still control all manned space operations, and there are limits to what an independent organization can achieve without the support or permission of a government agency. It is, therefore, in an organizations' interest to foster a relationship with NASA or other national agencies. Of course, NASA benefits from these relationships too, able to take advantage of private sector innovation. The involvement of private sector collaborators has created space for more fashion-conscious approaches to spacesuit design (Siceloff, 2017a). Boeing, for example, having been contracted by NASA to transport NASA crew to the ISS, have developed the "Boeing Blue" spacesuit for its crew (see Figure 3.3). Boeing's

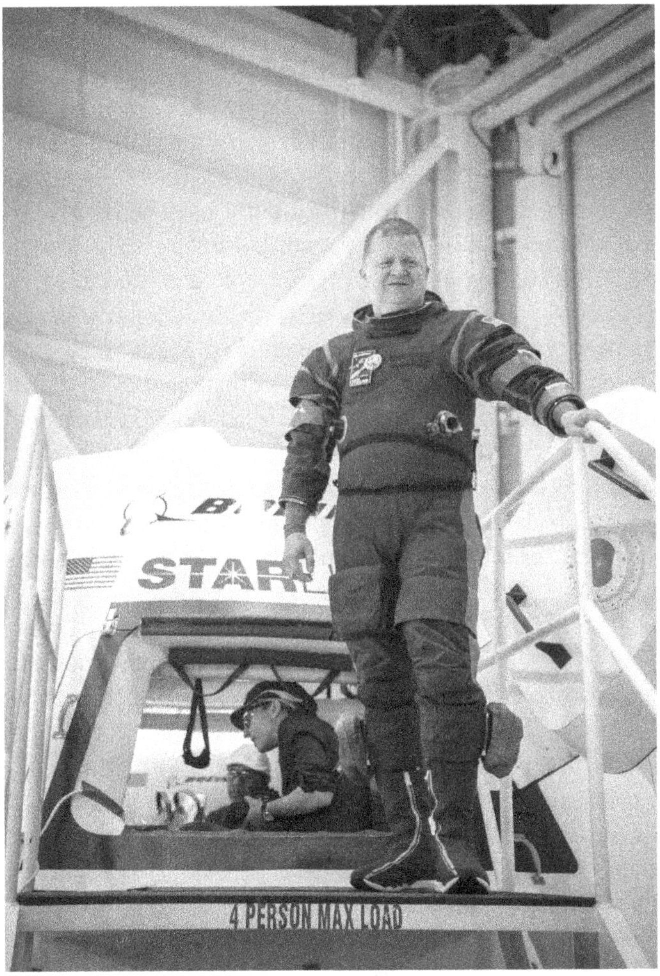

Figure 3.3 NASA astronaut Eric Boe wears Boeing's blue spacesuit designed to be worn by astronauts flying on the CST-100 Starliner for NASA's Commercial Crew Program. The suit is lighter and more flexible than previous spacesuits but retains the ability to pressurize in an emergency. The boots, designed by Reebok, give the suit a sportswear feel. Astronauts will wear the suit throughout the launch and ascent into orbit as well as on the way back to Earth. © Boeing/NASA

suits are "lighter, more formfitting" than NASA's own flight suits (Eric Boe, cited in Wall, 2017). Associations with activewear are explicit in the sneaker-like boots designed by Reebok, which Boeing describes as "akin to cross trainers." Boeing presents the suit as "tailored for Starliner," electing to refer to tailoring rather than engineering and, in doing so, drawing attention to the suit's aesthetic features (Siceloff, 2017b). Such designs may perhaps provide early indications of the kind of flight suits that could be worn by future space tourists during short-haul spaceflights offered by Boeing and their competitors.

Suiting the space tourist

Astronauts' activities, responsibilities, and, therefore, their suits, will differ significantly from those worn by space tourists. Astronauts' spacesuits need to accommodate a vast range of different environments and activities. The design, and thus the experience of wearing, differs from suit to suit. There is no "universal spacesuit" (Frost, 2015). The numerous different missions and conditions that require spacesuits each have unique needs. EVAs range from spacewalks to moonwalks, and may eventually include activities on the surface of Mars. As the length of space journeys increases, and with it the variety of extraterrestrial activities that may be carried out, there will be an increasing need for flexibility is spacesuit design (Jordan, Saleh, and Newman, 2006, p. 1135). Astronauts on spacewalks or on the surface of other planets will be required to carry out maintenance and experiments, and so mobility will continue to be a primary concern. With these activities often lasting six hours or more, suits must "accommodate all physiological needs, including hydration and waste functions" (Cadogan, 2015). Further, astronaut's spacesuits are designed for users with specialist knowledge and training (Gast and Moore, 2011). These complex and varied spacesuits for NASA, other government space agencies, and perhaps astronauts privately employed by commercial spaceflight operators, will have very different functional requirements from those worn by space tourists for leisure activity.

The spacesuit needs of space tourists will significantly diverge from those of astronauts and space explorers. While space exploration missions will likely see increasing demand for suits worn during EVA on the surface of asteroids or other planets, space tourists will not initially be undertaking EVAs on the surface of alien planets. For this reason, space tourism operators will initially only need spacesuits for short spacewalks or for emergency protection, if at all (Cadogan, 2015). When emergency suits are required, operational demands will be very different from those of other spacesuits. An emergency suit for commercial passengers will be required to do little more than protect its wearer from depressurization. Existing emergency space suits, such as the David Clark Company's Contingency Hypobaric Astronaut Protective Suit, CHAPS), serve the sole purpose of protecting the wearer from loss of cabin pressure and hypothermia (Messier, 2011). These pressure suits require no pockets or appendages to carry tools, no specially developed joints to enable mechanical operations, and no monitoring equipment. As a result, the cost compared to a fully equipped EVA spacesuit is significantly lower, and the potential for designers to focus on aesthetics over functionality are increased. Smart materials, and the simple requirements of an emergency pressure suit, mean that they may have "the comfort and aesthetics that conventional spacesuits cannot have" (Stylios, 2015).

At present, predictions differ as to whether future space tourists will be required to don pressure suits for their flights. David Ashford (2002, pp. 62–63) has suggested that any future space tourists on board suborbital flights may need to be dressed at all times for the possibility of depressurization. He suggests, therefore, that space tourists will need to be allocated a simplified version of the pressure suit worn by astronauts, which is "loose fitting and reasonably comfortable" with a "visor that closes automatically to seal the suit if cabin pressure is lost." Since passengers will not operate equipment, these suits, when pressurized, do not need to allow the degree of mobility that is so desirable for astronauts. Emergency pressure suits on board a space hotel may be even more simplified than CHAPS. In space hotels, though there may be freedom of dress for the duration of the stay, Ashford (2002, p. 63) suggests that the emergency pressure suits stored for all guests could be simple spheres, much like the Personal Rescue Enclosures proposed by NASA in the 1970s (Pearson, 1976, p. 67).

George Stylios (2015) proposes that most commercial spacesuits will be the equivalent to the oxygen masks on commercial airlines, deployed only in an emergency. The same comparison is made by the Civil Aviation Authority (2014, p. 283) in a UK-government review of commercial spaceplane certification and operations, where they note that "pressure suits with their own oxygen supply would provide an alternative method of protection" to "drop-down masks [that] would not be effective above 30,000 feet cabin altitude." The classification of such suits as emergency equipment is likely to mean that neither comfort nor aesthetics will be foremost in the mind of designers. Government regulatory agencies have not yet introduced a policy requiring emergency suits to be available to commercial spaceflight passengers (Webber, 2017, p. 144), and unless this regulation is introduced, the demand for commercial pressure suits may be limited. Stylios predicts that commercial spaceflight operators will be more likely to invest in prevention, concentrating on the safety of their spacecraft rather than on the unlikely eventuality that one will fail.

If the Civil Aviation Authority and equivalent bodies choose not to make the wearing of a pressure suit a requirement, feeling that safety systems are robust enough to eliminate the need for additional precautions, commercial spaceflight operators may elect not to offer suits to their customers. This decision would offer commercial spaceflight passengers the freedom to wear more comfortable or attractive clothing during their flights. Existing practices suggest, however, that these passengers will not have complete freedom to wear whatever they choose, as spaceflight operators' own safety policies will continue to determine what passengers will wear on board. Parallels can be drawn with the regulations that govern the choice of clothes for parabolic flights. While fireproofing and pressurization are less of a concern on board parabolic flights, providers have other safety concerns, arising from the lack of motion control that is experienced in weightlessness, and the likelihood of collision with fellow passengers. The Zero Gravity Corporation requires the majority of its participants to wear Zero-G branded flight suits—one-piece zip-up garments that are not dissimilar to those worn by NASA astronauts during training. On rare occasions, passengers are given special dispensation to wear their own clothes. In such cases, the corporation approves clothing on a case-by-case basis. Their approval is dependent on two key safety requirements: that the garment must not obstruct the passenger's vision, and that it should not "get in the way of other participants" (Abiera, 2016). Future spacewear designers would be wise to follow these same guidelines.

This ability for commercial spaceflight operators to determine, to an extent, their own safety policies, grants companies such as SpaceX and Virgin Galactic opportunities to provide flight suits for their passengers that are sleek, comfortable, and perhaps most crucially for some, visibly branded. Virgin Galactic have proposed branded coveralls, with rubber-soled socks "to help people maneuver around in weightlessness" (Klotz, 2012). Much like the current operators of parabolic flights, they propose that all passengers should wear the same suit, emblazoned with the provider's logo and designed with the intention of expressing a visual identity that Virgin Galactic pilot David Mackay describes as "very Virgin" (Klotz, 2012). These branded garments will have souvenir value (there will be a limited edition jacket for the first passengers), marking wearers as members of a very exclusive club (Alvarez, 2016). Branded flight suits are not dissimilar to those worn by elite pilots and astronauts in training, and seem to present the wearer as an honorary astronaut. Rather than giving passengers the freedom to wear their own clothing, Mackay expresses the view that "a lot of people actually do want to look like an astronaut when they go into space" (Klotz, 2012), and branded flight suits are one way to achieve that appearance.

Other commercial operators, whose need to attract customers is at the heart of their activities, are similarly invested in developing spacewear that is aesthetically pleasing. Virgin Galactic has employed Y-3, an Adidas brand overseen by Japanese fashion designer Yohji Yamamoto, to develop flight clothing for its crew (Alvarez, 2016). The collaboration with Y-3 "will eventually extend to garments for crew and passengers." The prototype is a one-piece black flight suit, 3D-engineered from heat-resistant Nomex, donned via a long zipper that runs from the collar down the chest, terminating at the waist. Zipped pockets run around both calves, adjacent to additional zippers that run to the top of the shin (Howarth, 2016). Lawrence Midwood, senior designer at Adidas Y-3, stated that the flight suit design was driven by functional concerns, his principle concerns being safety and “the ability to move freely around the cockpit” (Kleanthous, 2016). Closely resembling existing NASA and USAF one-piece flight suits in silhouette, the design follows established approaches. Features including the black color and leather detailing adhere to the established aesthetic of Y-3 and Yohji Yamamoto, and so stylistic concerns are evident, but the principal appeal of the suit is its association with a fashion brand.

It seems that in flight suit design it is difficult to escape the influence of the Space Age, which itself derived its flight suit design principles from the Jet Age (Gooden, 2012, p. 65). This continued dependence on the zip-up, single color, form-fitting one-piece as the default form for a flight suit is both logical and intuitive, but its continued use into the new commercial space age should not be taken for granted. There is room for critical reappraisal of the form-fitting one-piece and its presumed necessity. Arguments for retaining this silhouette stem from its proven practicality. Over more than half a century, the one-piece flight suit has served its purpose. With safety and practicality in mind, other aspects of recently designed flight suits have also tended to imitate features previously seen on NASA spacesuits. A case in point is Angella Mackey’s (2014) modular neoprene flight suit, designed for Spaceport Sweden and Smart Textiles, a sleek and “hyper-functional” figure-hugging, two-piece suit, that features wrap-around pockets that can be relocated from arm to leg, and vice versa, or removed entirely when safety demands a smoother silhouette. Mackey’s design takes account of the Earth-based and space-based activities that commercial space travelers will encounter, including wind-tunnel and g-force training. For that reason, many of the features of her garments respond to reports by those who have experienced space training (but have not yet visited space), including padding around the shoulders to protect from the tight seat belt worn during g-force training. Mackey has followed NASA’s lead in adopting modularity for flexibility and close fit for enhanced safety, and in doing so has produced a design with an unmistakably Space Age feel.

It may not be the case, however, that such designers are imitating the Space Age aesthetic by default. Commentary surrounding commercial space travel highlights designers’ desire to consciously engage with the history of space travel while looking to the future. Given the still significant achievements of the first Space Age, and the continuing presence of the Space Age aesthetic, it seems that design of the new commercial space age must reference this history. The winning Z-2 Technology suit was described by NASA (2014d) as “pay[ing] homage to spacesuit achievements of the past while incorporating subtle elements of the future.” Similarly, Adam Wells, head of design at Virgin

Galactic, says of his collaboration with Y-3, that "it places what we're doing in the context of history." Wells refers to this history not as contained solely in the past, but as ongoing, presenting the work of Virgin Galactic as history in the making. Wells feels that Virgin Galactic is "a forward-looking enterprise that is looking to define the future on our own terms." His collaborator on the flight suit, Lawrence Midwood, agrees that one of their principal goals was to create "a sense of Virgin Galactic continuing mankind's journey . . . to broaden our horizons on Earth and potentially beyond" (Kleanthous, 2016). Describing his design of the Y-3 flight suit, Midwood quotes Yohji Yamamoto: "With one eye on the past, I walk backwards into the future" (Kleanthous, 2016), acknowledging that designers can no more peer into the future than the rest of us, and can only aspire to a futuristic aesthetic, looking to the past and present to predict what that future might hold.

SpaceX's solution to the problem of designing for an uncertain future is to turn to science fiction and fantasy for inspiration. One of Elon Musk's stated goals is to develop a flight suit that looks "like a 21st century spacesuit" (Cuthbertson, 2016). His statement suggests a desire to distance his products from those produced during the first Space Age, and instead to be instrumental in establishing the design aesthetic of the next space age. Musk has sought spacesuit designs from Hollywood costume and prop designers, Jose Fernandez, after Fernandez's work on suits for fictional superheroes Iron Man and Batman, and Chris Gilman, whose portfolio also includes the founding of spacesuit design firm Orbital Outfitters (Cuthbertson, 2016; Oberhaus, 2015a). His decision to collaborate with costume designers rather than fashion designers reflects Musk's intention to create a suit with a science-fiction aesthetic, and to position his customers as heroes of the new space age. By publicly embracing this relationship to science fiction and fantasy, commercial space entrepreneurs acknowledge that "fictional technology . . . sets audience expectations" (Shedroff and Noessel, 2012, p. 6), and that their customers may expect developments in technology and design that mirror those they have seen in science fiction. Space tourism and the colonization of distant planets have, over the past century, been presented to audiences as science fiction, and so it is inevitable that the expectations of the first space tourists will be shaped in part by fiction and fantasy.[4]

SpaceX's approach is the polar opposite of Virgin Galactic's. While Virgin Galactic prioritize functionality, SpaceX have made style their principal concern, to the extent that functional elements needed to be retrofitted to Fernandez's suit. Musk describes the need to compromise in order to achieve spacesuit designs that are both functional and stylish. When he unveiled the SpaceX suit on Instagram in 2017, he wrote that it "was incredibly hard to balance aesthetics and function. Easy to do either separately." The result is a form-fitting two-piece in white and black, with charcoal piping. During the early stages of the design process, style took precedence over utility, as Fernandez worked for six months on the aesthetics of his suit before engineers were brought in to reverse-engineer a design that would be "functional for flight" (Boyle, 2016). Musk chose not to unveil the suit to the public until it had been pressure tested, recognizing that earlier prototypes would have lacked credibility if they had not been proven ready for use or, indeed, if the design had been too dissimilar to the suits already in use by NASA. It is noteworthy that this suit more closely resembles an Apollo spacesuit than other commercially developed flight suits, thanks largely to its white color and the shoulder-mounted Stars and Stripes. These similarities to the Apollo suit are indicative of SpaceX's intention to present their planned commercial missions as a continuation of NASA's Apollo Missions, and thereby locate the suit, and SpaceX, within a history of pioneering technical innovation.

NASA, SpaceX, and Virgin Galactic have each taken a different approach to their spacewear and spacesuit prototypes but, nonetheless, similarities are emerging that will help to establish the aesthetic of future space travel. Many of the similarities between the various commercially developed flight suits are a consequence of functional requirements, particularly in the case of pressure suits, where the need for an airtight container dictates, not least, the need to cover the entire body. It is no surprise that the stylistic features of commercial flight suits borrow from sportswear, given that the activewear market is similarly driven by technological innovation and practical requirements. As in many design fields, flight suit design is an exercise in problem solving that begins with identification of the major functional requirements. However, as is also the case in other design fields, functional requirements cannot fully justify the duplication of central design features, and should not result in derivative design. There is, even when meeting functional requirements, opportunity for aesthetic difference. Further opportunities for designers to differentiate their spacewear from others' emerge when we acknowledge that the same functional requirements are not common to all space travel environments. As the next section of this chapter will show, spacewear designers will have more freedom for stylistic innovation when focusing their attention on routine wear for IVA, where there are fewer technical concerns to govern their approach to design.

Off-the-shelf and out-of-this-world

For tourists staying on board a space hotel, or partaking in long-duration spaceflights, a more extensive wardrobe will be required than a single flight suit. NASA differentiates between crew members that are "suited," protected against the environment of extravehicular space, and "unsuited," "wearing the type of clothing that . . . might be worn on Earth" (NASA Technical Standard, 2015, pp. 194–195). Astronauts who stay on board the ISS have an assortment of clothing available to be worn while they are "unsuited," but this clothing has not yet been subject to as much scrutiny or development as the spacesuit, in part because off-the-shelf clothing has proved adequate for wearing in intravehicular conditions. The use of commercially available clothing on board the ISS is made possible because the conditions on board are closer to those on Earth than those of extravehicular space. On board the ISS there is "ambient air," replicating Earth's breathable atmosphere, and comfortable temperatures are maintained in order that astronauts can breathe and dress as they would on Earth. Routine wear does not need to be pressurized, nor does it need to protect from extreme temperatures or radiation, as the exterior shell of the ship, combined with the carefully controlled atmosphere of the cabin, offers the necessary protection. It is because of the Earth-like conditions of the cabin that that the intravehicular environment offers most potential for the wearing of fashion items, designed by future fashion designers rather than spacesuit engineers. The "routine wear" worn in the cabin, distinct from "suits" that are "designed to protect the crew member from differences in environment, such as pressure, atmosphere, acceleration, or temperature" (NASA Technical Standard, 2015, p. 194), can be evaluated in terms that are already familiar to Earthwear designers.

Routine wear includes the clothing worn for everyday activities on board space vessels, not for the extraordinary EVAs that occur outside the craft in open space (see Figure 3.4). Lisa Fratto (2005) suggests that it was not until the twenty-first century, prompted by the advent of the Mars program, that NASA began to recognize intravehicular clothing as a "critical element of space exploration," while Annalisa Dominoni (2005, p. 13), head of research programs at SpaceLab, attests that an overhaul of IVA garments could directly contribute to "improved living and working conditions in space." Twenty-first-century studies have begun to identify the need for garments that are not only safe but "attractive" too. Both Fratto (2005, p. 37) and Dominoni (2005, p. 9) have called for IVA garments that cater to ISS crew members' personal tastes. Dominoni (2005, p. 9) calls for the provision of a more diverse selection of garments, not only to create a more professional environment but, perhaps more importantly, in order to improve the well-being of ISS crew. Fratto (2005, p. 39) argues for the importance of appearance for crew morale, given the effects of isolation on mood. In this heavy stress environment, crew members can become physically and emotionally sensitive, requiring garments that are comfortable for the body and pleasing to the eye (Dominoni, 2005, p. 9).

NASA's Space Flight Human-System Standard lists the requirement of clothes to be worn on board spacecraft and habitats. It guarantees each astronaut clothes for his or her own exclusive use, that are "clean and durable," "non-toxic and flame resistant," "comfortable in fit and composition and compatible with the environment, e.g., temperature and humidity, in which it will be worn," and designed for donning and doffing without assistance (NASA Technical Standard,

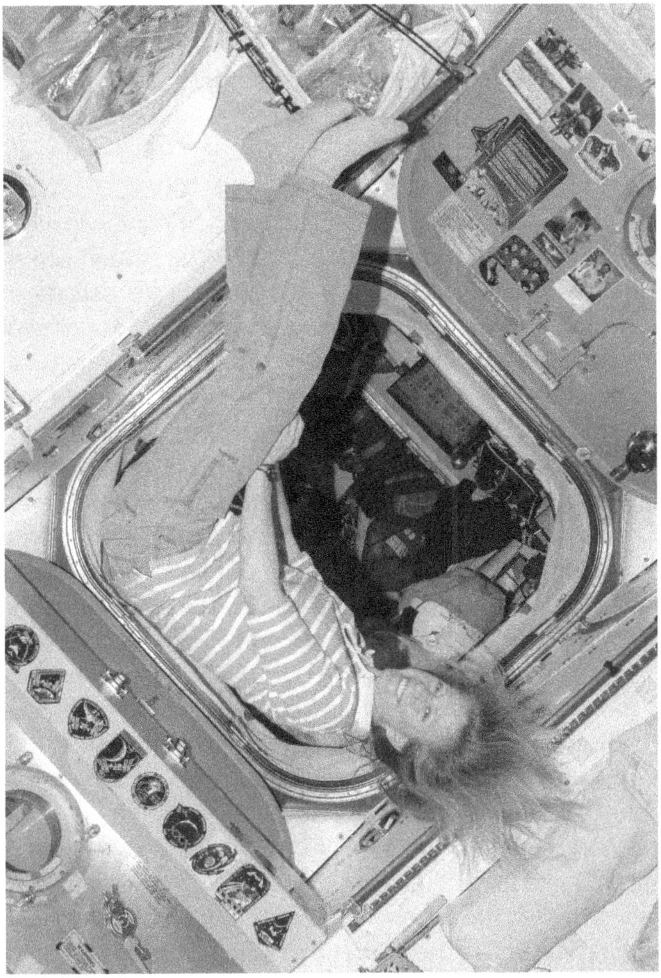

Figure 3.4 NASA astronaut Karen Nyberg on board the ISS. She is wearing COTS (Commercial Off-The-Shelf) polo shirt and cargo pants by outdoor outfitters Cabela's. © NASA

2015, pp. 92–93). These requirements seem surprisingly few, given the tightly controlled environment of spacecraft and space stations. However, they must also be considered in conjunction with stowage requirements and mass restrictions. With limited transport space on board shuttles and supply modules, and on board space stations, only limited storage space can be devoted to clothes. For the ISS, garments are selected in part for their mass, with 250 g allowance for T-shirts, 350 g for cargo shorts, 650 g for cargo pants, and 80 g for a pair of socks. The total mass allowed by each astronaut over a year is 75 kg, meaning that each garment must last a rather long time before replacement. T-shirts and socks are replaced once a week, long-sleeve shorts every fifteen days, and cargo pants every thirty days. Clothing supplies are delivered incrementally to the ISS by cargo module (Broyan, Schlesinger, and Ewert, 2016, p. 2). Once a garment has been worn to its limits, disposal burden becomes a concern. Worn clothes are disposed of in nonreusable spacecraft, which are deorbited, and break up in Earth's atmosphere (Dunbar and Petty, 2003).

Aside from mass, key factors in selecting routine wear for space stations are flammability and hygiene (Orndoff, 2015b, p. 27). The oxygen-rich environment of a space station, together with the lack of opportunity for escape, makes fire a significant risk. Fabric must be slow burning, and must not melt; it should also be antistatic "to reduce the risk of sparks" (Fratto, 2005, p. 38). Cotton has historically been the fabric of choice, due to its low flammability in contrast to man-made fibers. Concerns about the risk of fire have also resulted in a requirement for route wear to be "low linting" (Fratto, 2005, p. 40). In a weightless environment, lint from clothes and other fabrics can float into filters, causing them to become clogged. Cotton garments are shaved and washed before being packed for transport, in order to minimize lint. The length of time that an item of clothing must be worn on board a space station, with no laundry facilities, necessitates a concern for hygiene (Broyan, Schlesinger, and Ewert, 2016, p. 2). As they will be worn for two hours of exercise each day, and for fifteen consecutive days, it is essential that textiles for exercise wear be resistance to odor-causing germs (Orndoff, 2013). NASA has initiated studies into the bacteria-inhibiting properties of a range of materials, to test the suitability of commercially produced fabrics and garments.

After tests on a range of off-the-shelf garments, involving astronauts and cosmonauts on board the ISS, Evelyne Orndoff, NASA's senior textiles engineer, and her team have shortlisted a range of textiles with properties that are desirable for long-duration missions. The cotton that was ubiquitous in clothing for the Space Shuttle program, on board the Mir, and recently the ISS, is discarded in favor of lighter alternatives that produce less lint (Orndoff, 2016). Merino wool, which was previously declared too uncomfortable due to its rough texture, is now more appropriate thanks to commercial developments that have resulted in a smoother fiber. While wool and cotton have similarly low flammability, wool has the advantages of being capable of regulating temperature more effectively, having fine weave for lower mass, and having no odor when dry (Meggs, 2015). For shorts, Orndoff and colleagues have selected "the lightest polyester that we can buy on the market" [*sic*]. Orndoff (2013) has proposed ultrafine merino wool for exercise shirts, lightweight polyester for exercise shorts, and modacrylic shirts for routine wear, all commercially available. Crew on board the ISS are allocated T-shirts and socks, manufactured by Noble Biomaterials from antimicrobial silver-coated fibers, and available to commercial consumers under the brand X-static (NASA, 2007).

Crew members are permitted a limited number of "crew preference" garments (Orndoff, 2015b, p. 9). These are commercially available "off-the-shelf" (COTS) garments, from various commercial manufacturers, that have been selected for inclusion in a NASA catalogue (Fratto, 2005, p. 2; Broyan, Schlesinger, and Ewert, 2016, p. 3). For its catalogue, NASA selects commercially available clothing that meets its safety standards and mass restrictions, that is, for example, long- and short-sleeve shirts with a mass of less than 550 g and 450 g respectively (Fratto, 2005, p. 18; Orndoff, 2015b, p. 9).[5] Following practices established by other military organizations, NASA turns to COTS products in order to save time and money in the development of specialist products. Products can be obtained more cheaply and quickly when the cost and time invested in development has occurred at the expense of an external vendor. Though it is acknowledged that "custom clothing would better meet requirements," the development of specialist IVA garments "would drive up costs and limit availability" (Fratto, 2005, p. 18).

NASA's procurement of off-the-shelf products is not always as straightforward as making selections from product catalogues. There is some flexibility in the definition of COTS, as used by the US military organizations in general. The US Air Force defines COTS as including commercially available products with some minor adaptions, and commercial products that have not yet been made available to the civilian market (AFMC, 1993, p. 10). Garments with commercial off-the-shelf origins can be customized by NASA so that they are better suited to the extraterrestrial environment. For example, flame-retardant or antibacterial chemical finishes can be applied to a garment after purchase, mission logos are embroidered on the chest, and Velcro strips are often sewn on for the temporary attachment of pockets and tools (Fratto, 2005, pp. 41–42, see Figure 3.5). However, there is only a limited amount that can be done to match off-the-shelf garments with the requirements of space travel. NASA have no input into the design of off-the-shelf products, nor can they "control changes to that design" (AFMC, 1993, p. 10).

The use of off-the-shelf clothing requires market surveys to be conducted before appropriate garments can be shortlisted, and tests to be carried out in order to assess the suitability of each shortlisted garment for the intravehicular environment. While manufacturers may have thoroughly tested their products for a number of extreme conditions, the unique conditions of the cabin of a space station necessitate additional tests. They are, for example, tested for heat transfer and moisture transfer to determine their flammability and hygiene properties (Fratto, 2005, p. 37). For all COTS equipment, including clothes as well as off-the-shelf electronics and other mass-manufactured products, NASA testers are particularly concerned with long-term reliability. The number of times that an IVA garment must be worn, without the possibility of replacement, means that there must be some compromise. More robust fabrics that are strong and

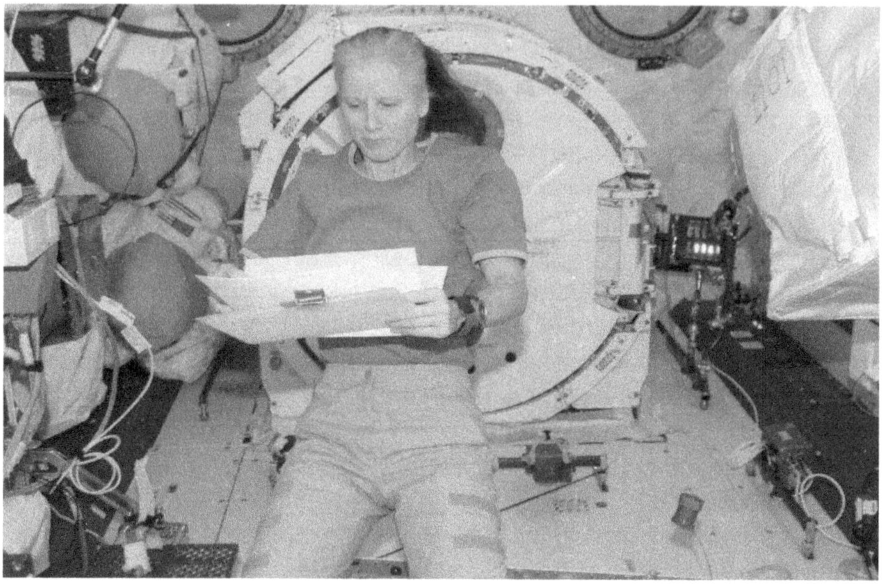

Figure 3.5 NASA astronaut Shannon Walker, on board the ISS wearing a COTS T-shirt and cargo pants modified with Velcro strips that can be used to attach pockets and tools. This image also illustrates the effect of weightlessness on the T-shirt collar that is described in more detail in the next chapter. © NASA

abrasion resistant may sometimes be preferable even when they are heavier (requiring more stowage space than a weaker, lightweight alternative).

While NASA is still dependent on commercially available "off-the-shelf" wardrobes for much IVA, government and commercial space agencies recognize that routine garments are not fully compatible with microgravity. In studies of images broadcast from the ISS, Annalisa Dominoni (2005, p. 4), observes "collars rucked up around [crewmembers'] necks" and "backs uncovered because garments have not been designed to consider [neutral] body postures" (see Chapter 4). The effect, she writes, is "slovenliness," not befitting of representatives of the human race. The unsuitability of some COTS garments for microgravity is generally not known during testing, as there is "no practical way" to test the effects of weightlessness on a clothing (Fratto, 2005, p. 6). Parabolic flights do offer some limited opportunities for testing, but they do not allow multiple tests to be combined. For example, they do not enable the testing of garment reliability and performance during a long microgravity exercise routine, nor do they enable observations of how bacteria will multiply on fabric over several days of weightlessness (Kemsley, 2013). Parabolic flights are also inadequate for testing the form and fit of garments for long-term space travel. During such short periods of weightlessness (20–30 seconds per parabola), passengers are typically in motion, and garments are not given the time to settle around the body, nor is the body itself given time to settle into its neutral position. Furthermore, the bodies of passengers are not yet distorted by the effects of long-term weightlessness. As a result, the behavior of clothing in the microgravity conditions of a space station are unpredictable, and inadequacies in COTS clothing are sometimes only revealed when they are worn long-term by space station crew.

Dominoni's observations were made as part of an Italian Space Agency–funded study into the suitability of existing routine wear for activities on board the ISS, leading to the development and testing of a new IVA clothing system, VEST. The study considers the requirements of IVA clothing, and identifies "countermeasures" that can be taken to offset the effects of microgravity on the human body (Dominoni, 2003, p. 280). With an emphasis on flexibility, the VEST project proposes garments with several forms of elasticity. Elastic sleeves and trouser legs, and strings that allow manual adjustment of the waist, will make it possible to accommodate the changes in body shape that crew members experience over a long stay on the ISS (Dominoni, 2005, p. 8). Pockets will not be fixed, rather, there will be separate pouches that could be attached to Velcro so that they could be repositioned to account for the different needs and reach of crewmembers (p. 11). The addition of pouches is among the features that were informed by Dominoni's focus on the practical needs of working astronauts. Dominoni (2005, p. 13) refers to her VEST project as a work of industrial design, not fashion design, and it is important that this distinction be made when contrasting the requirements of astronauts and other spaceflight participants.

The flexibility of the VEST clothing system extends to its suitability for different gravitational conditions. Dominoni (2003, p. 283) remarks on the importance of designing garments that can be worn on Earth as well as in space, so that crew can wear them during training. Her study finds that familiar clothing helps crew to adjust to life on board the ISS (p. 282). Similar considerations will be vital for any future spacewear designer, as a requirement to design for both microgravity and 1G would significantly influence design decisions. Unless accommodations are made for changing on board commercial spaceflights, it is likely that short-term space tourists will necessarily wear the same clothing on the ground as in orbit. Anticipating the short-duration suborbital flights that are proposed by operators including Virgin Galactic, it seems likely that space tourists will need to wear their chosen spacewear on Earth as well as in microgravity, and so their garments cannot be impractical in a 1G environment.

Studies such as those by Dominoni and Fratto focus on the needs of inhabitants of the ISS, but their findings may also be applicable for future missions beyond Earth's orbit. Evelyne Orndoff (2015a) has declared that "a new vision is needed for the development of a wardrobe for NASA's journey to Mars." In anticipation of long-distance space travel, NASA has embarked on a program of "logistics reduction," focusing in particular on reducing the mass of crew clothing. Orndoff

(2015b, pp. 10; 12) has been tasked with developing a clothing system with reduced mass and volume for wearing on board Orion and, eventually, inside a Martian habitat. NASA will not be able to continue its established cycle of using, discarding, and resupplying for long-duration missions, as this would place a heavy logistical burden on the mission (Meggs, 2015). Clothing will no longer be treated as a consumable, as long-duration missions will not have the benefit of resupply modules to periodically replenish a crew's wardrobe. The challenges posed by long-duration missions are addressed in Orndoff's *Space Wear Vision* (2015b). Orndoff's study focuses on the need to reduce the volume and mass of clothing by addressing a number of related concerns. By reducing the mass of individual items, while also increasing long-term wearability so that fewer garments are needed, the overall mass of clothes to be supplied can be reduced. Orndoff's investigation has so far led to, among other things, "cargo bags that can be reconfigured for crew outfitting" (Broyan, Schlesinger, and Ewert, 2016, p. 1).

It is interesting to note that Orndoff's (2015b, p. 4). presentation of the *Space Wear Vision* project begins by establishing that "ideal" spacewear is, for NASA, unachievable. She describes a method that involves imagining the ideal, and then stripping back that vision until reaching an achievable compromise that strikes a "balance between the ideal and the [available] resources." Ideal spacewear is presented as a dream of the distant future and for now, she proposes, we can only hope to begin to bridge the gap between the inadequate IVA clothing of today and the garments that may one day be perfected by future generations of spacewear designers. For the time being, her task is to select textiles and garments from "the best that's available now," in the commercial market, rather than to develop new clothing technologies (Meggs, 2015). She relies on advances made by other commercial organizations, describing, for example, advances made in the wool industry in the smoothing of fibers.

If, as Orndoff suggests, ideal spacewear is unachievable for NASA, might the emergence of a commercial spaceflight industry create the right conditions for ideal spacewear to be developed by other organizations? The need for compromise arises because commercially available off-the-shelf garments are not intended as spacewear. NASA is required to relentlessly test and treat garments before use, and to accept imperfect fit, because commercial suppliers have not yet begun to design everyday garments for microgravity. Typically, COTS garments are originally intended for Earth activities and conditions, with most designed for sports or outdoor activities. *Space Wear Vision*'s self-imposed requirement for compromise evidences the need for a shift in attitude, not away from commercial suppliers, but toward commercial suppliers with an explicit interest in the space sector, who understand the microgravity environment. At present, no such clothing company exists. Orndoff acknowledges NASA's historical use of purpose-made specialist garments for all space travel, and the shift to commercial off-the-shelf clothing in recent decades (Meggs, 2015). While this approach may be less costly, it will continue to force compromise until the commercial fashion industry is equipped to supply more suitable garments. Once commercial organizations begin to develop garments that are specifically intended for the weightless environment, government space agencies may be able to eliminate the need to carry out their own tests, and instead rely on those commercial organizations to conduct the tests as part of their own commercial development processes. If fashion designers and apparel manufacturers follow the lead of product designers such as the Cosmic Lifestyle Corporation (see Chapter 1) or large commercial space ventures such as SpaceX, developing their own commercially available garments for space, they will have opportunities to forge relationships with NASA and other government space agencies as well as future space tourists and commercial space travel providers.

In pursuit of the ideal spacewear, some future spacewear designers may find that it is advantageous to maintain distance from NASA and other large spaceflight operators. NASA has established a precedent for dressing astronauts in branded, identikit flight suits, and early indicators are that commercial spaceflight providers will do the same. NASA's selection of COTS clothing, while offering some choice to astronauts, is limited in shape and color. David Mackay's assertion that Virgin Galactic passengers will want to look like astronauts is, argues Mark Timmons, unrealistic, reflecting a lack of awareness about consumers' need for personal expression through fashion. Timmons sees the problem as arising from industry that is still dominated by engineers. "They're all techno geeks," he says in conversation with the author. "We as designers have a massive opportunity to challenge that mindset." As a designer, "once you remove gravity you can do really interesting and challenging things with the [clothes] that you wear," and both designers and consumers will seek something extraordinary when they consider dressing for this new environment. Timmons argues that fashion designers like to engage with what is possible, and has never been done before, not what has been tried and tested. "You don't go into space to look backward," says Timmons, "you go into space to look forward."

4

THE CLOTHED BODY IN MICROGRAVITY

Perhaps the most important thing that future spacewear designers will need to understand is how the relationship between clothes and the body is altered by weightlessness. The body in space is not the same as the body on Earth. When designing for the weightless body, one must take account of shape, posture, and capacity for movement, all of which are affected by gravitational changes. Fluid moves so that mass is repositioned, deforming the body and, over time, muscle atrophy leads to thinning of the limbs. Without the force of gravity, the natural tensegrity of the skeleton and muscles takes over, and the body adopts a neutral posture. Everyday actions such as standing or walking are not possible, so clothing must accommodate new sets of gestures and actions. Movement must be relearned, or substituted for different gestures that have been designed for the weightless environment. The body, its shape and size, and everything it does, are altered in ways that must be accommodated by spacewear designers.

Weightlessness can be presented as "freedom from gravity" (Bureaud, 2006), but with this freedom comes a loss of control that affects movement and interaction with objects including wearables. This loss of control arises in part as a direct result of weightlessness, but can also be a consequence of restrictive wearables. Spacesuits for EVA have contributed to audiences' understanding of how weightlessness affects human motion, with artificially generated pressure and stiffness that restricts the wearer's movement and gives moonwalking astronauts a distinctive "hop." Everyday movements and gestures that our bodies perform under Earth's gravitational conditions are very different from those that are possible or desirable in space, requiring that astronauts undergo training to "translate" familiar gestures (Gast and Moore, 2011, p. 323). The astronaut's "action possibilities" differ depending on what he or she wears.

The dialogue between garment, space, and body is very different for rigid and flexible materials, and for different gravitational conditions. The hard shells of some spacesuits and the soft fabrics that can be worn for IVA respond differently to varying gravitational conditions. In this context, rigidity and flexibility can be defined as the extent to which a material is distorted by gravity. While a rigid structure tends to maintain its shape regardless of gravitational conditions, soft, flexible fabric becomes malformed by its own weight in Earth's gravity. This malformation, commonly referred to as drape, is an essential consideration in fashion design, informing choice of fabric and cut. Drape is, however, currently understood as the extent to which a fabric resists gravity (Cusick, 1965; Cadigan, 2014, p. 140). Future spacewear designers will need to significantly revise their understanding of drape behavior when designing for microgravity. Engineers and manufacturers of the rigid spacesuit concern themselves with enabling the maximum possible flexibility that will enable astronauts to carry out the gestures that are need to perform EVA. By contrast, designers of spacewear for intravehicular activity sometimes have to contend with too much flexibility, and the potential for uncontrollable distortions of form, as when clothing floats upward and outward, away from the body.

"One small step": Movement in microgravity

Those who have experienced weightlessness describe a loss of awareness of where their body ends and where its surroundings begin. Without the sensation of pressure on the epidermis, skin "stop[s] playing the role of sensor between the 'interior' and the 'exterior,' between 'me' and 'what is not me'" (Bureaud, 2006). If the "external limits" of the body are unclear, then so too is the division between body and clothes. This sensation is compounded within a spacesuit, in which the internal workings of the body are externalized. The externally worn spacesuit becomes involved in actions that normally take place inside the body, and transits related data to onboard computers. In this environment, the act of wearing is not a matter of swathing the body in cloth but, rather, blurring the divide between body and space.

The phenomenology of microgravity can be affected as much by the wearing of a spacesuit as it is by the state of weightlessness. Immersion in a spacesuit makes the wearer hyperaware of his or her own physicality. Astronaut Chris Hadfield describes "the gentle sound of the fan and the liquid cooling ventilation" that acts as a constant reminder that a spacesuit is working to sustain the flow of oxygen to the body (Savage, 2015). When the fan "shuts off it becomes very quiet. [Astronauts] can hear their own body noises (breathing, swallowing, pulse)" (Frost, 2012). On a spacewalk, recalls Hadfield, the sound of "your own breathing" is "your [constant] companion" (Savage, 2015). These sounds are characterized by their closeness. In contrast to the vast expanse of space, where the horizon is so incredibly distant that it curves around the entire circumference of a planet, everything that the astronaut hears is generated from within inches of the astronaut's body. The vast and distant visual landscape bears no relation to the relatively minute and close aural landscape. The absence of external sound dissociates the astronaut from actions and events outside of the suit, including actions carried out by the astronaut himself. NASA astronaut Piers Sellers describes the "odd" sensation of banging "a hammer or metal tool and bang[ing] it against something . . . hear[ing] absolutely nothing" (Watts, 2011). Without the aural feedback that we take for granted within Earth's atmosphere, the astronaut is reliant on his visual and haptic confirmation that he has carried out an action as intended. Thus, events do not provoke a complete sensory experience, and the astronaut must become ultra-aware of visual cues, including those that appear to contradict what his other senses are telling him.

A spacesuit so significantly inhibits familiar interactions with other objects, and weightlessness so significantly alters the character of those interactions, that familiar gestures must be relearned in preparation for space travel. The design of a spacesuit dictates the quality and extent of its wearer's movement, to the extent that motion can be said to be choreographed by the designers and engineers who have developed the suit and its components. The awkward "hop" of the Apollo Crew on their moon landing was, writes Nick Taras (2014), "less to do with the Moon's gravity and more a result of the lack of maneuverability of the spacesuit." The rigidity of the pressurized suit restricts movement of joints. This joint stiffness is in part a result of the limited flexibility of the outer layers and in part a consequence of the compression of gases inside the suit. Joint movement is so significantly restricted that it impairs the wearer's ability to perform everyday tasks. The suited astronaut cannot fully extend his arms or legs, nor fully rotate his shoulders or wrists, nor manipulate his fingers as easily as he can without pressurized gloves and, as a result, maintenance tasks are physically difficult to perform. The distance that a suited astronaut can reach is known as the "reach envelope," and is significantly smaller than the distance that he or she would be able to reach when unsuited. Within the reach envelope lies the smaller "work envelope"—the area within which the astronaut can reliably operate. The work envelope is limited by the range of motion in joints; the length of the astronaut's limbs; the astronaut's strength, which dictates the extent to which the joint can bend; and the area of visibility, which is limited by the helmet (Schmidt, 2001, p. 174). All of these physical restrictions, imposed by the wearing of the suit, affect the speed at which the astronaut moves and works, as well as cause significant physical strain on the body.

Describing her many experiences on board parabolic flights, Susan Buckle of the UK Space Agency tells me how, even without a spacesuit, "your body acts very differently [in microgravity]. It doesn't really know what to do." Thus, astronauts must learn to perform familiar actions, such as turning a wrench, in conditions that defamiliarize these everyday actions to such as extent that they must be learned anew. Astronauts must learn to adjust their movements and gestures according to the range of motion that is possible in each of their various spacesuits. In training, astronauts learn a process of "translation" by which they become accustomed to "action possibilities" (Gast and Moore, 2011, p. 323). Annalisa Dominoni (2015, loc. 289; 275) proposes that these gestures must not simply be relearnt, but redesigned, and that designers of products and tools for space must consider "not just objects but the gestures and uses people might make of them." The process of designing tools and wearables to be used by astronauts in microgravity necessarily involves designing the gestures that permit their use. The design of gestures and the design of objects must go hand in hand, and one will dictate the form of the other. For example, the handle of a tool must be designed in such a way that it can be grasped through a pressurized glove, and the way in which the astronaut is trained to grip that tool will depend on the design of the handle. In the case of a suit or other wearable, new ways of walking or transporting the body must be designed, taking account of the restrictive design of the spacesuit and, in turn, the spacesuit must be moderated to enable this movement to occur. This is an iterative process, in which the work envelope, and the gestures that it permits, inform developments in the design of the tools or spacesuit, and vice versa.

The process of relearning and gestures involves becoming accustomed to the lack of resistance. Earth's gravity forces the body and its parts to move in largely two-dimensional ways, working with or against gravity. In general, upward movement requires greater exertion than downward movement. When one lifts an arm upward, the muscles resist gravity, and when the arm is relaxed, it lowers, carried downward by the force of gravity. However, without gravity, we are "liberated" to move freely in "three-dimensions." In a weightless body, "no [more] muscular force is required to move" in any one direction than another (Bureaud, 2003). This freedom of movement can be seen as liberating, with gravity seen as a limiting, restricting force. However, as mankind has historically designed all of its actions and tools for use in Earth gravity, everyday activities are more often enabled by gravity than restricted by it. Familiar movements and gestures have evolved to work with gravity, and in some cases are impossible without it.

Having experienced weightlessness on no fewer than fourteen parabolic flights, choreographer Kitsou Dubois has found this freedom to be a reminder of how many of our everyday movements are dependent on gravity and contact with a stable surface. So many of the movements that our bodies perform on Earth originate from an "impulse point," at which two opposing forces trigger movement. Walking, for example, requires the gravitational pull of the Earth to attract the body toward the ground with each step, and the stable surface of the ground is needed as a surface from which to launch the next step. A gesture such as twirling in a dress also requires the presence of a stable surface from which the body can propel itself. Once this motion has started, gravity and friction are required to end it. As Dubois experienced first-hand, a body spinning in microgravity will continue to spin, even in a perpetually relaxed state. The freedom of weightlessness is therefore a blessing and a curse: to gain freedom it is necessary to sacrifice control. This problem is illustrated in everyday gestures and activities, such as donning and doffing (see Chapter 3).

The difficulties associated with controlling movement in microgravity arise in part from the lack of stability that affects both the weightless body and the objects with which the space traveler must interact. On Earth we are rooted to the ground, and our fashion reflects this. Shoes are the interface between body and ground, and mediate our experience of groundedness. They are objects through which we experience the effects of gravity, both weighing our feet downward and cushioning the weight of our bodies. The shape of shoes reflects an approach to design that is informed by gravity and its effects. Shoe designers concern themselves with stability and balance, clearly differentiating the sole—the base through which the body is weighted to the Earth—from the upper. Shoe design begins with the assumption that the shoe will be weighted to a horizontal surface, and this platform becomes the starting point for the rest of the design.[1] The fact that weightless objects float rather than fall "obviates the need for . . . flat surfaces" (McKinnon, 2015), rendering the sole of the shoe, at least as we know it on Earth, virtually useless. Without gravity weighing the body to a surface, the need for shoes as we know them on Earth is eliminated. Susan Buckle notes that "when you are floating, your feet become very unimportant." Like her fellow passengers on board Novespace parabolic flights, she wears white trainers during her weightless experiences, but only so that she can fall feetfirst on the floor of the aircraft when gravity returns to the cabin. Astronauts on board the International Space Station (ISS) do not typically wear shoes, as there is neither a floor on which to stand, nor the possibility of maintaining a standing posture without aid.

The inability to stabilize the body by simply adopting a standing posture means that other methods of stabilization have needed to be sought for weightless environments. On board the ISS, stabilizing holds exist on all surfaces, and are as readily grasped with the hands as they are used with the feet. Thus, astronauts are grounded at whatever point their body makes contact with the solid interior of the cabin. Astronauts can be seen propelling themselves through the cabin with their hands, much like a child swinging along monkey bars (NASA, 2015a). While the interior of the ISS does have footholds incorporated into its surfaces, by hooking their feet into a hold astronauts sense pressure on the top, not the soles of their feet. When astronauts need to remain at a workstation, they must anchor themselves in place by means other than sitting or standing. Anchors can come in the form of foot braces, consisting of one or more adjustable straps, through which the astronaut may slide his or her foot. Less formally, astronauts can be seen hooking their toes under handrails (Behrendsen, 2013). The main point of contact, therefore, is not the sole of the foot, but the top of the forefoot.

Astronaut Scott Kelly (2016) reports developing callouses on the top of his feet, while his soles grew "baby soft" through disuse. Were astronauts to wear shoes while using these braces and rails, it would be the toe box, not the sole, through which the body is grounded.

In order to establish a connection between feet and surface, astronauts require not just shoes, but shoe systems. Space stations have employed a range of foot restraints in order that astronauts may hold themselves in place while freeing their hands to work. These are generally in the form of straps or interlocking components, with a foot piece that locks into a platform. Astronauts on the Skylab space station wore soft lace-up boots with rigid soles, to which were screwed triangular cleats (see Figure 4.1). These cleats matched a grid of equilateral triangles cut from an aluminum plate that lined several of Skylab's interior surfaces. These grids served to stabilize the astronaut, working "as both a handhold and a locking surface for the triangle shoe" (Chowdhury 2016). In order to remain stationary for a long duration, an astronaut could rotate his shoes to lock the triangular cleat in place by locking it into a triangular cavity (Watkins and Dunne, 2015, p. 343). Although Skylab's cleat system did have advantages, it has not been used in more recent space stations, where straps and bars offer the freedom to more quickly relocate from one foothold to another, without the need for specialist footwear.

Andreas Vogler's (2005, p. 1) "Design Study for Astronaut's Workstation" reveals astronauts' dissatisfaction with the foot restraint as a means of stabilization, partly due to the fact that the standing position is not so easily

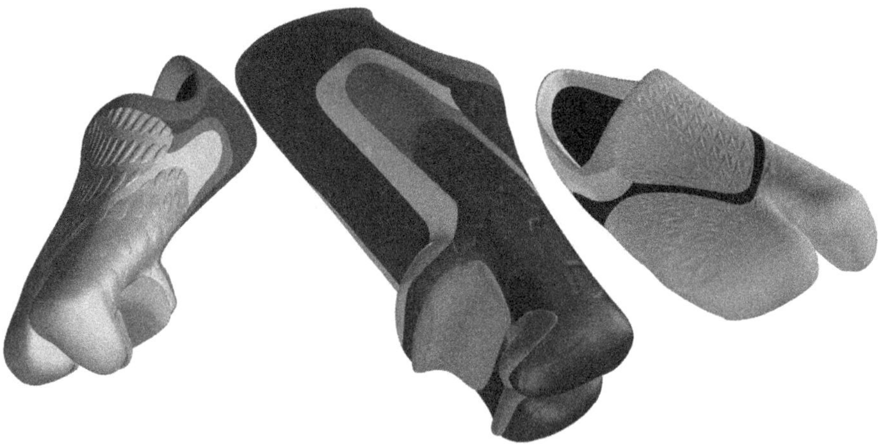

Figure 4.1 Edward Howell's proposed design for Zero Gravity Prehensile Footwear. It enhances the grabbing capabilities of the toes, enabling astronauts to carry objects with their feet while they use their hands to navigate through the cabin of a space station. This illustration depicts one of a series of design by Howell that have a variety of appendages and protrusions for different grasping and holding tasks. © 2010 by Edward L. Howell

adopted or maintained in microgravity as it is on Earth. A standing position is stable in Earth's gravity, and this tradition of associating the standing position with stability has informed the use of foot restraints at workstations on board the ISS, as well as during extravehicular repairs. However, remarks Vogler, the standing position only provides stability when the body is affected by a subjectively vertical force of gravity. Standing and walking, on Earth allow "a very wide reach and good employment of our muscles for physical work," and require constant physical effort from the body's core muscles. By contrast, "lying is the preferred mode for relaxing and sleeping, since muscular support of all members is minimized" (p. 1). In a weightless environment, these two "restraint modes" are indistinguishable. The body may only adopt a position similar to standing or lying by being tethered to a stable object. Rather than using the core muscles that are put to work in a standing position on Earth's surface, foot restraints place constant strain on the shin and toes (Behrendsen, 2013, p. 15).

Vogler (2005, p. 1) argues for stabilization closer to the body's center of gravity. On board parabolic flights, photographers and instructors are stabilized with tethers attached to their waists, so that they can remain rooted to the floor of the plane and able to assist free-floating passengers. A waist restraint affords "good body control," but also significantly limits an astronaut's range of motion, and hence their ability to carry out required tasks (p. 2). Vogler argues that a "chair-like restraint" may offer the best solution, tethering the body at the thighs so that the user may extend his or her reach beyond what would be possible with a waist restraint (p. 3). Tethers or bars located at the thigh suit the neutral posture that the body adopts in microgravity, thereby minimizing the physical strain that is required to maintain a stable position. Such restrains exist in the form of the Munich Space Chair (Igenbergs, Naumann, and Pfeiffer, 1996) and for use of waste collection systems (NASA, 2002). Clothing designed to allow stabilization close to the body's center, as Vogler proposes, might allow or even facilitate physical connections between the waist or hips and a nearby surface, just as shoes are designed to act as the interface between the body and the surface of the Earth.

Feet are, of course, used not only for standing and stabilizing the body on Earth, but also for transporting the body from one location to another. Weightlessness denies the possibility of walking or running, and as a result astronauts' feet no longer play a role in transporting their bodies around the spacecraft. Susan Buckle describes how "it takes [astronauts] a long time to figure out how to move around the space station . . . They go head first as if they're diving through the different modules, so [they] would never walk around like on Earth." Footage of astronauts navigating through the modules of the ISS reveals that they travel using their hands, not feet. They move through the cabin by propelling themselves between handrails, so that the "primary movement" is performed by the "fingers hands, arms, and shoulders" (Behrendsen, 2013,

p. 13). Hands become the main point of contact with a solid surface, and so the sensation of groundedness originates from the hands, not feet. One might, therefore, argue the case for gloves that adopt some of the features of footwear.

These alternative methods of grounding the body introduce the possibility of designing without the need for a stable platform at the soles of the feet. Space travelers may find shoes redundant, or may desire footwear with other practical or aesthetic functions that are unrelated to walking or standing. Designer Edward L. Howell (2011) has recognized the potential for footwear that enables feet to be involved in actions other than stabilizing the body. His Zero Gravity Prehensile Footwear has hooks, extrusions, and valleys above and below the toes and the ball of the foot. These protrusions extend the functional ability of the toes, enabling grasping and holding, so that the wearer may make equal use of hands and feet when navigating through a cabin, or holding and transporting tools (Howell, 2017, p. 8). The design of the interior of the ISS has already led to a kind of substitution of hands for feet, in so far as astronauts use their hands to navigate through the interior. So that feet are not rendered unimportant, Howell's footwear enables the interchangeability of hands and feet for the carrying out of tasks in microgravity.

For future spacewear designers, the floating, weightless body introduces new possibilities not only for footwear design, but also for the design of garments that might otherwise terminate at the ankle. It is worth noting the extent to which the shape and length of Earth garments are determined by the need to permit standing or walking. Dresses and skirts, for example, must be designed to allow free movement at the hem, and less at the waist, leading to variations on a bell-shaped silhouette. Designers discarding the need for a stable platform at the feet may also find themselves redesigning garments around the whole lower part of the body, freed from the need to permit walking or standing. At the same time, the need to stabilize the weightless body introduces constraints that are not present in the Earthwear design process. Spacewear must allow for tethers, bars, or other stabilizing attachments to be drawn around, or connected to, the body, ideally close to its center of gravity, at the thighs, hips, or waist. Such clothing will enable kinds of physical connections between the body and its environment that are taken for granted on Earth as the body is pulled downward onto the surface on which it stands.

Posture, shape, and the weightless body

The human body is so shaped by the gravitational conditions in which we grow and live that, on entering different gravitational conditions, our bodies immediately begin to reshape. Reduced gravity effects the body so significantly in so many ways that there are, argues Dominoni (2003, p. 284), "new parameters of wearability" that arise from the microgravity environment. For astronauts, clothing designs must accommodate changing body shapes, a variety of functional needs specific to the activities that take place on board spacecraft, as well as aesthetic concerns arising from the poor fit of garments that have been designed to fit a body shaped by Earth gravity. Weightlessness has short-term and long-term effects on the human body, leading to changes in body shape and size that must be accommodated by the spacewear designer. In the short term, fluids are more evenly redistributed so that the upper part of the body increases in volume, leading to a broader chest and shoulders. Astronauts who spend a long time on board the ISS experience loss of muscle and bone mass, as they weaken from disuse. As leg muscles atrophy, the legs become slimmer, contributing to the top-heavy appearance of astronauts who have been on board the ISS for a long-duration mission (see Figure 4.2). In order to minimize the long-term effects of weightlessness, astronauts must exercise for two hours per day in the onboard gym, tethered to a treadmill via a harness, or using equipment that artificially generates resistance using piston-driven vacuum cylinders (Loehr et al., 2011). These immediate and delayed effects of gravity on the shape of the body mean that off-the-shelf clothing selected to fit the body of an astronaut on Earth will not fit so well in space. When worn on a journey to space, preselected clothing will be almost immediately tighter on the upper body, as fluids redistribute, and will become gradually looser on the lower body, as muscles atrophy.

Wearables play a role in partially offsetting the negative consequences of microgravity, helping to preserve bone density by applying pressure to the body in ways that imitate the effects of gravity. When not wearing their off-the-shelf clothing, astronauts on board the ISS may be found wearing gravity-loading suits, designed to counteract the effects of long-term weightlessness. The Russian Penguin (or "Pingvin") suit houses a series of bungee cords running from the shoulder to heel of the garment, which apply axial force to the body, stimulating the gravity force that we experience when we live on the ground to "prevent the loss of bone density [and] muscle strength" (see Figure 4.3). Stirrups can be tightened to "increase the axial load on [the wearer's] leg" (JAXA, 2014). Although this does not make the body any less weightless (the wearer may still float about inside the space station), it does counteract the effects of weightlessness on the body. The latest generation of Gravity Loading Countermeasure Skinsuit (GLCS) simulates gravitational loading with a complex web of fibers of varying elasticity "that simulates the loading of terrestrial gravity in a more continuous, shoulder-to-ankle manner" than older suits lined with bungee cords (Murray, Waldie, and Newman, 2014). These suits have echoes of the girdle-like pressure garments that resulted from the relationship between the fashion and spaceflight industries that emerged in the 1960s as shapewear manufacturers found themselves developing pressure suits for high-altitude flights (see Chapter 1).

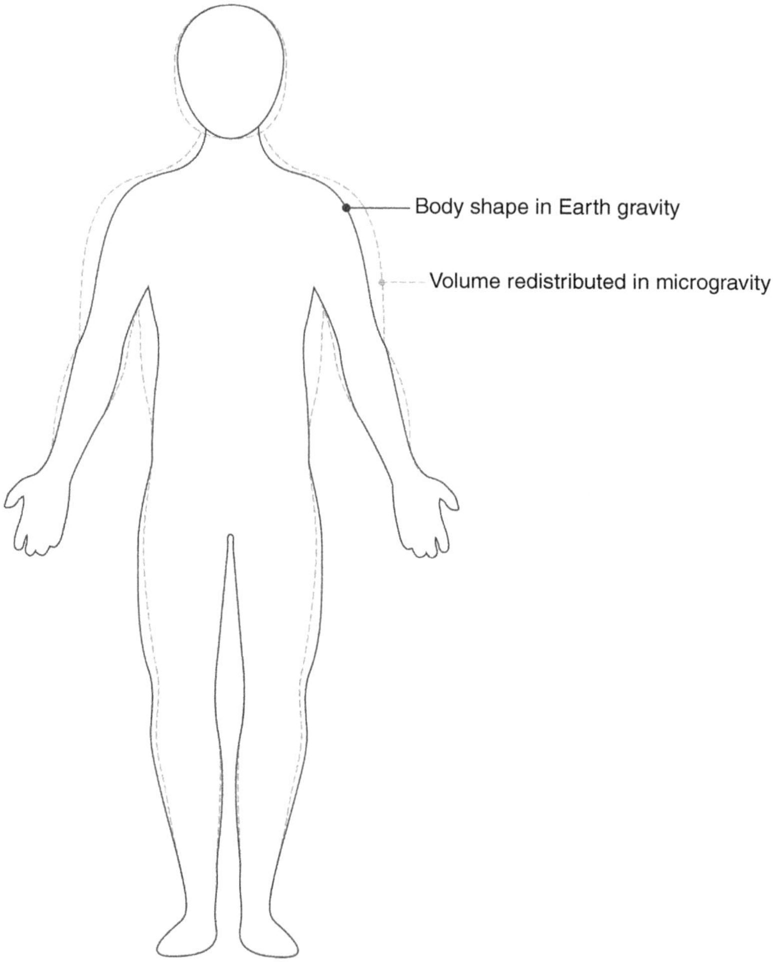

Figure 4.2 The body changes shape as a result of the effects of weightlessness. The upper part of the body increases in volume, leading to broader chest and shoulders, as fluid is evenly dispersed around the body and, after a longer term, the legs become thinner as muscles atrophy

The most recent developments in gravity loading clothing move beyond simply compressing the body from shoulder to toe, moving toward the simulation of a personal gravitational environment. In 2011, NASA commissioned the Draper Laboratory to develop a suit that stabilizes the astronaut's body, so that movements may be carried out in microgravity. Their experiments resulted in the V2 Variable Vector Countermeasure Suit, which employs gyroscopes to track the body's movements, and flywheels to create artificial resistance to replicate the sensation of gravity. The suit restores the sense of orientation that is lost in microgravity, by generating resistance when movement is made parallel to a

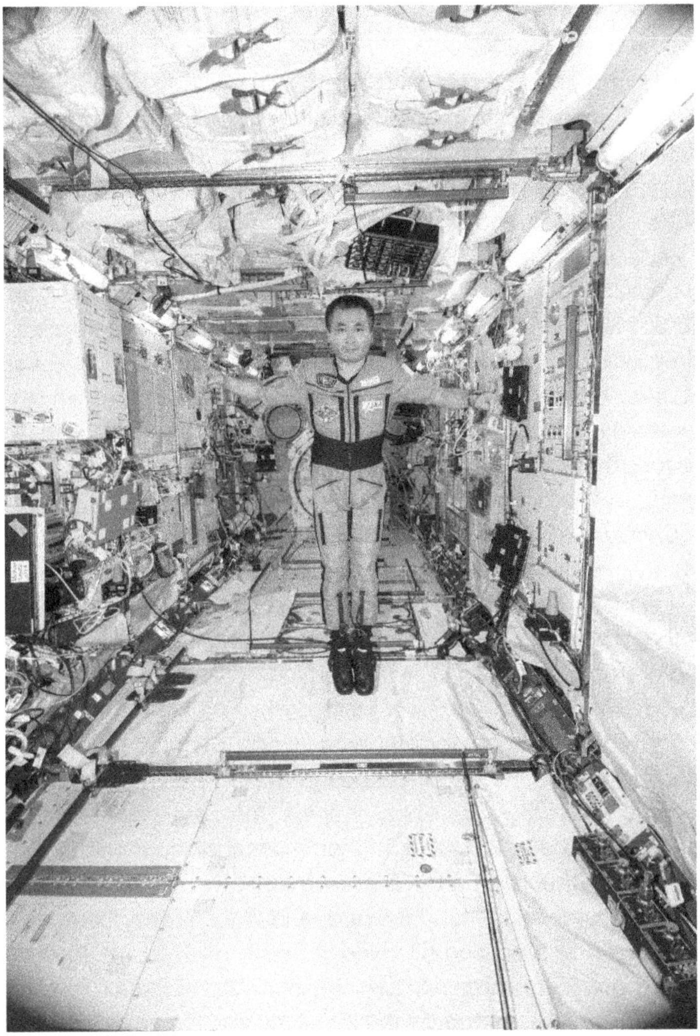

Figure 4.3 Japanese astronaut Koichi Wakata is pictured wearing the Penguin-3 antigravity pressure/stress suit on board the ISS. While most of the bungee cords are concealed within the suit, this image shows the stirrups that hold the bottom of bungee cords to the heels, from where they stretch up the body to the shoulders, thereby simulating the load of gravity on the shoulders and spine. © NASA

downward direction, and less or no resistance when movement is perpendicular to that direction (Duda, 2014, p. 3). The same technology may lessen the threats imposed by gravity on Earth's surface. Engineers at the Draper Laboratory recognize Earth-based applications for the V2 suit, where it can provide "gait or movement stabilization for the elderly, or rehabilitating individuals." The suit could be "programmed to provide a kinematic envelope of least resistance during walking," or counterbalance that would reduce the likelihood of falling (Duda, 2014, p. 4). In this way, wearers on Earth would no longer be slaves to gravity.

Such developments in wearable technology contribute to the potential for the emergence of the post-gravity human. An advanced stabilizing pressure garment that isolates its wearer from the effects of gravity (or indeed, of the absence of gravity) may enable the human species to transcend concerns about those effects. This resistance and stabilization requires that a downward direction is "arbitrarily specified" (Duda, 2014, p. 3). The post-gravity human is, in this way, centered on self. When orientation is subjective, "you are your own anchorage" (Bureaud, 2006). Individuals determine their own planes of reference. With no natural downward force, the direction of up and down must be defined by the manufacturers or users of the V2 suit. The realization that engineers must artificially allocate upward and downward directions—concepts that are so ingrained in the human psyche that they once seemed natural or universal—contributes to a kind of awakening, similar to the one experienced by the Arts Catalyst philosophers and performers who reflected on their experiences of weightlessness, and who came to a shared understanding that orientation is a subjective experience (see Chapter 1). If future generations can define and redefine upward and downward however they choose, orientation might one day be considered a naïve and antiquated concept that limited the imagination of earthbound ancestors (Doule, 2014a, p. 93).

As well as changes in anatomical proportions, perhaps the most significant factor affecting the wearability of clothes in microgravity is posture. When not making an effort to extend the limbs or spine, the weightless body relaxes into a "neutral posture" that is less straight than the standing, or anatomical, posture one might adopt when on Earth. The astronaut's neutral posture is much like that of a snowboarder, with head forward, and limbs slightly bent, as if partway between standing and seating (Dominoni, 2003, p. 279, see Figure 4.4). The posture is not identical in all cases; a variety of similar postures are observable in recordings of astronauts on board the ISS: typically, the arms float forward and outward, and are slightly bent at the elbow; legs are gently bent at the knees, even when feet are hooked under a rail to stabilize the body; hips are slightly bent so that the thighs slope forward, and the head is held slightly forward of its usual position. While being interviewed, astronauts are often seen clasping their hands together in front of their waist in order that their arms do not float outward and obstruct the view of their coworkers. Since there is no force pulling the body in any particular direction, this same posture is adopted regardless of the astronaut's orientation in relation to the cabin or other objects. One consequence of the neutral posture is ill-fitting garments, as can be observed in images of astronauts wearing routine wear in microgravity environments. It is this posture that causes shirt hems to rise, exposing astronauts' lower backs, and collars are caught uncomfortably under the chin (Dominoni, 2005, p. 4).

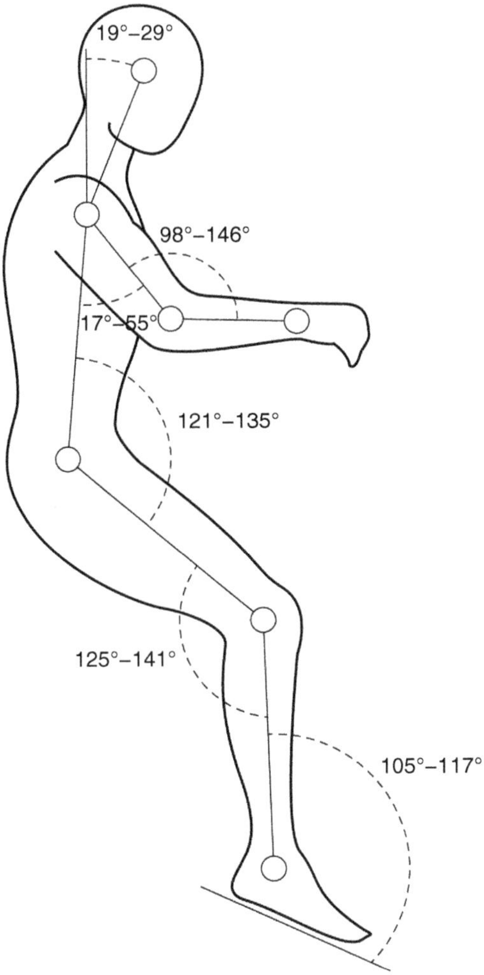

Figure 4.4 In microgravity, the body adopts a neutral body posture. The head and shoulders tilt forward, arms float, slightly bent, in front of the chest, and knees are bent. Adapted from Whitmore et al. (2013, p. 10, fig. 2), originally from data collected during observations of twelve Skylab crew members

Clothes, including the COTS garments worn on board the ISS, are currently designed for an upright pose known as the "anatomical position," with spine erect, legs straight, and feet flat on the floor (Watkins and Dunne, 2015, p. 36). In the anatomical position, the head faces forward, and the head, pelvis, and feet sit approximately in vertical alignment. The body in this alignment stands perpendicular to the ground, and experiences gravity through vertical compression. This is the pose of the dressmaker's dummy, but very different from the pose of a human body in microgravity. When not subject to the force of gravity, and therefore not forced toward the ground, the body naturally falls out of this alignment. Limbs relax, and pressure is relieved from the spine, and the head tilts forward. Dressmaker's dummies are today, as standard, designed in the anatomical position. There are adjustable dummies that can be expanded to cater to different body shapes and sizes, but not to alter the dummy's posture into a neutral posture.

Observing the fit of garments designed for the anatomical position on a body with neutral posture, it is possible to identity points of concern where the fabric is stretched uncomfortably tight, falls untidily loose, or fails to adequately cover the body. Particular points of concern for dressmakers and designers are the collar, armscye, elbow, waist, seat, crotch, and knee. The neutral posture requires garments that are longer in the back; shorter in the front; looser around the knees, seat, and shoulders; and with neckhole and waistband positioned further forward than on garments designed for 1G. At the waist, there is potential for gape at the back and unwanted pressure at the front. The hemlines of shirts lift at the back, and rear of the waistbands of trousers and shorts gape or are pulled downward, leaving an area of the lower back exposed. Meanwhile, the front of a waistband can become rucked and tight, applying uncomfortable pressure to the abdomen. A longer seat seam at the rear, and shorter rise at the front, helps position the waistband on a horizontal line.

Dressmakers who work with dummies in the anatomical position may be ill-equipped to imagine solutions to the problems that arise when clothing the body in neutral posture. Images taken on board the ISS can be useful in helping spacewear designers to identify points at which a routine garment fails to adequately fit when the body rests in a neutral position. Dressmakers dummies that are posed neutrally would be essential in the design and manufacture of spacewear, and would help designers to test possible solutions to the problem of poor fit. It is also helpful to look at the pattern-cutting techniques that are already in use by tailors catering to stooped and seated postures, which introduce many of the same problems as the neutral posture. While in the era of mass-market fashion the anatomical position is taken for granted, historical pattern-cutting resources provide valuable insight into how one might accommodate alternative postures just as readily as manufacturers currently accommodate different dress sizes. Spacewear designers might learn from Compaing and Devere's *Tailor's Guide* of 1855, which includes notes on the consideration of posture as well as size and shape (Breward, 2016, p. 22). Compaing and Devere (1855, p. 5) note that, just as a man may be "stout" or "thin," he may also "stoop forward or lean backward." They describe the attitude of a stooping man, which must be accommodated by cutting a shorter front and a longer, wider back, as well as a narrower chest, since a curved back is likely to be accompanied by shoulders that hunch forward (pp. 9–10). Further adjustments are required for a stooping body that bends forward (p. 11). Plates illustrate how, to accommodate a forward bend, the waist seam of a coat must be raised at the front and lowered the back (plate 6). These poses—the stooped posture, bending forward—share similar features with the posture of the upper part of the weightless body. A much more recent guide, Winifred Aldrich's (2015, p. 217) *Metric Pattern Cutting for Women's Wear* describes adjustments that need to be made for "figure problems" including stooping posture. Aldrich's diagram of a stooping silhouette, showing waistline and hemline that angle upward toward the rear of the figure rather than resting horizontally as they do on an upright figure, illustrates problems similar to those observed in the fit of clothes worn on board the ISS.

Earthbound spacewear designers might also observe the clothing of seated wearers to understand the consequences of bent postures. In particular, it may be useful to look at how clothing for wheelchair users is designed to accommodate the seated position, which causes similar stresses and rucks as the neutral posture. The design of clothing for wheelchair users is "based on data obtained for users in a sitting position" (Nowack, 2001, p. 878), prompting "adjustment of the bottom part of clothing in length and cut" and "removal of excess fabric in front" (Nowack, 1999, p. 1343). Brands including IZ Adaptive manufacture garments that meet these requirements, including pants with additional room in the seat so that the waistband rests horizontally on a seated wearer, and tight fit on the underside of the knee (Lubitz, 2016). As in spacewear design, clothing design for wheelchair users is concerned with garments that can be removed with minimal assistance, and is therefore often donned and doffed differently to clothing designed for the anatomical position (Wang et al., 2014, p. 551). Garments may, for example, be divided into a number of separate pieces so that they are disassembled during undressing, a little like a spacesuit. It would be advantageous for future designers to be familiar with such approaches so that they may be applied in the context of spacewear design.

Drape in weightlessness

Designing garments for space, one must consider not only the changes to the body, but also changes in the behavior of fabric. The effects of gravity on the body are, in part, experienced through the clothes that we wear. When cloth rests against our skin we are reminded not only of the presence of our clothes, but also the presence of gravity. The relationship between clothes and the body is governed by the shape and posture of the body, the cut of the garment, and the behavior of the fabric from which it is constructed. None of these can be considered in isolation. Nor can they be considered without an understanding that every one of these aspects of clothing design is affected by gravity. On Earth's surface, garment design involves seeking a balance between drape and structure: that is, the extent to which areas of the garment respond to or resist gravity (Cadigan, 2014, p. 140). In a microgravity environment, this relationship between drape and structure is thrown out of balance, and the behaviors that we have come to expect from certain fabrics are no longer reflective of qualities (such as weight) that play such an important role in the design of Earthwear.

The weight of a fabric is of particular importance in the design of Earth fashion. Indeed, fabrics are categorized according to their weight, using a measurement of ounces per yard or grams per meter (gsm). A 100 gsm cotton, for example, may be classified as lightweight, while a 248 gsm cotton may be classified as heavyweight (Stecker, 1996, p. 211). These weights are also associated with the seasons, with lightweight fabric sometimes being referred to as "summer weight," and heavy fabrics as "winter weight." In these terms, the weight of a fabric is determined by its thickness and density, where density is usually determined by the tightness of the weave. On Earth, weight is closely associated with drape. The weight of a fabric dictates the way that a fabric will distort when draped over the body (Mei et al., 2015, p. 1; Hunter and Fan, 2008, p. 7). Fine, lightweight fabrics are selected for garments that should skim or cling to the surface of the body. Conversely, heavy fabrics tend to be less revealing, having the potential to disguise the contours of the body (Saville, 1999, p. 260). When a fabric is weightless, these rules no longer apply.

Microgravity will cause space tourists to experience the acts of dressing and wearing clothes in new ways. The behavior of fabric in microgravity is unpredictable, and dependent on many variables, not least the weight of the fabric. Even if one were to model the behaviors of one kind of fabric in microgravity, this model would not necessarily be applicable to fabrics with different properties, such as tighter weave or greater elasticity. Designers also need to consider that an understanding of the behavior of a certain textile in microgravity would not necessarily tell us about the behavior of a garment made of that same fabric. Garments do not behave in the same way as an untethered piece of fabric. In a garment, fabric is constrained, and the nature of that constraint depends on the design of the garment and the nature and behavior of the body that wears it. The deformation possibilities of any fabric depend on the ways in which it is cut, stitched, and worn. These and other variables would need to be taken into account over many years of microgravity experimentation before we could confidently predict, with any degree of accuracy, the way that a garment might hang on the body of a space tourist.

It is possible to see the effects of weightlessness on a limited range of clothing in recordings of astronauts on board the ISS. Thanks to an increase in the availability of parabolic flights, footage is now also available that shows the effects of weightlessness on a wider variety of fabrics and clothes, including, importantly, those that are designed to drape. Parabolic flights are being used for the authentic portrayal of weightlessness in science fiction and fantasy films, and in music videos (beginning with Ron Howard's *Apollo 13* in 1995 and, more recently, for the complete duration of OK Go's music video, *Upside Down and Inside Out*, 2014). Alex Kurtzman's *The Mummy* (2017) features a plane crash sequence for which a period of weightlessness was filmed on board a Novespace zero-g flight. The sequence shows Tom Cruise and his costar, Annabelle Wallis, floating through the cabin of a plane, grappling to strap on a parachute as they become weightless. Wallis, who wears layers including an unbuttoned jacket, is shown spinning through the cabin. As she floats weightlessly, the hem of her jacket lifts upward and outward, exposing her back, and forming a cape-like shape. Such stunt sequences, which are likely to become increasingly common as parabolic flights are on offer in a wider range of locations, provide insights into the way drape is affected by weightlessness and changing gravitational conditions. The sequence illustrates how garments that are designed to drape downward, hanging close to the wearer's skin, float away from the body in weightless conditions. Increasingly, film costume designers will need to be aware of the effects of weightlessness if they are to work on films that feature such sequences and, at the same time, fashion designers will be given access to an increasing archive of footage that provides insight into drape behavior in variable gravity conditions.

There are several problems encountered when attempting to consider the drape of fabric in a weightless environment. The first is that definitions of drape refer directly to gravity or weight. Of the numerous texts that explore and define drape, many explicitly or implicitly associate it with gravity: for Cusick (1965), drape is "a deformation of the fabric produced by gravity when only part of the fabric is directly supported"; Jiang, Cui, and Hu (2012, p. 661) describe how, "under the action of gravity . . . fabrics will droop . . . and form curved surfaces"; Cadigan (2014, p. 140) categorizes all fabrics according to the extent to which they "resist gravity"; and Cadigan (2014, p. 140) defines drape as "how fabric falls in space." Drape is measured by apparel and fabric manufacturers using apparatus known as drapemeters, which provide data about the extent to which unsupported fabric is pulled vertically downward by the force of gravity. The simplest measure of drape is the outward spread of fabric from a point of suspension. Some tests also measure the fabric waves, including the amplitude, curvature, and number of waves created when fabric is draped over a plate (Sanad, Cassidy, and Cheung, 2012, p. 354, see Figure 4.5). Such tests assume that gravity will exert a downward force on the fabric, causing it to fall or bend toward the Earth. When tests for the behavior and qualities of fabric rely so heavily on the effects of gravity, they are inappropriate for testing or defining fabrics for use in microgravity. A new approach is needed, one which measures and classifies fabrics and their behavior in weightless conditions.

The second problem arises due to the close relationship between drape, structure, and support in the design of garments. Erin Cadigan (2014, p. 140) presents drape and structure in direct opposition, as either falling or resisting

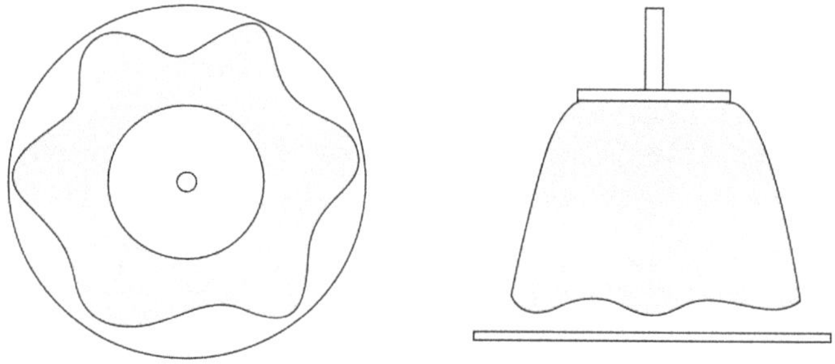

Figure 4.5 A drape test measures "how fabric falls in space" (Cadigan, 2014, p. 140). A circular fabric swatch is trapped centrally between two rigid plates with a smaller circumference, so that the edges of the fabric drape over the edge of the disk. A light source is placed above the center of the disk, so that a shadow is cast on a surface below the fabric. Drape is quantified by measuring the area of the cast shadow. Thicker, stiffer fabric tends to have a wide spread, as it forms a gentle curve from the edges of the plates, hence it casts a shadow with a large area. Finer fabric tends to have a smaller spread, as it hangs almost vertically downward from the edges of the plates, hence it casts a shadow with a small area. However, this is not the case in microgravity. Since drape behavior is so vital in dictating the form of many garments, and drape is a product of gravity, new methods of defining the behavior of fabric are required for microgravity environments

gravity. A fashion designer's goal, she argues, should be to pursue a "balance between drape and structure" (p. 125). This balance is not achieved only in the choice of textile, but also in the construction of the garment, which may be shaped so as to "fight . . . gravity" (p. 125). As illustrated in the drapemeter test, when drape occurs in Earth gravity, it involves the opposition of gravity and support, as one part of the fabric is held aloft by a supporting structure, and the unsupported parts of the fabric are free to drape downward. Texts that define drape frequently acknowledge that drape occurs only when "part of [a garment or cloth] is directly supported" (Hu, 2004, p. 54). Support works in opposition to gravity: gravity pulls the fabric down, while support holds it up. This support typically exists where the fabric comes into contact with the body, as it does at the shoulder seams and waistband. Earthbound dressmakers begin with the assumption that a garment will "hang" on a human figure.[2] Gravity causes a garment to rest on protruding parts of the body, so that fabric hangs on either side of the protrusion. There are points of direct support, typically located at or near to the top of a garment, at which gravity holds the garment to the body. On either side of those points of support, fabric may hang freely around the body (in a draped garment), or follow the contours of the body (in a form-fitting garment).[3] In a straight cut garment, side seams are perpendicular to the ground, so that, for example, the front and back hem of the garment both fall to an equal distance above the ground when the wearer is standing upright (Stecker, 1996, p. 199).[4]

Figure 4.6 indicates the points of contact with the body, from where the garment drapes in the gravity conditions of Earth's surface. In this example, in a men's polo shirt (similar to those worn by astronauts on board the ISS), the primary points of direct support lie at the shoulder seams. These primary points of contact, created by the pull of gravity, vary depending on the nature of the garment. Typically, primary points of contact exist where the body protrudes outward, creating a shelf that is parallel to the ground, or within about 45° of parallel. In a skirt, they are likely to be located on the hips. In a fitted dress, they might be located both at the shoulders and the hips. An artificial point of contact may also be created through elasticated seams or tight fit, such as an elasticated

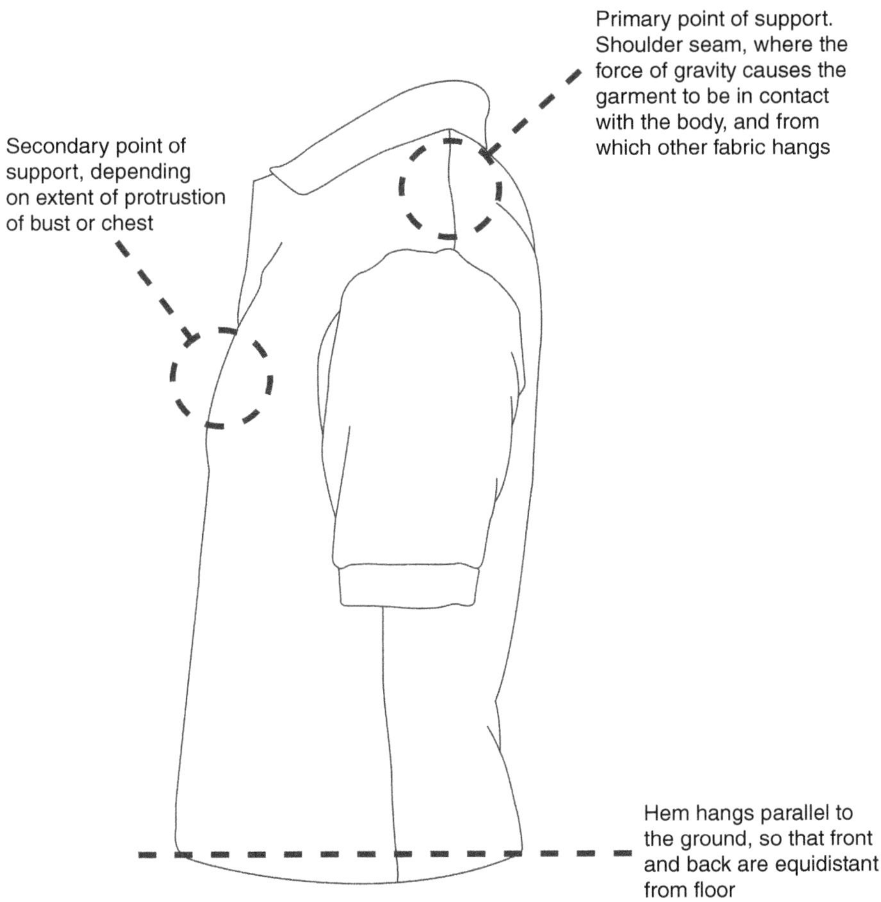

Figure 4.6 Diagram of an off-the-shelf polo shirt, showing primary and secondary points of direct support, at which the force of gravity causes the shirt to come into contact with the body. The remaining sections of the shirt are suspended from these points, so that the fabric hangs against the torso but is not supported by it

or drawstring waistband like those proposed in Dominoni's VEST project (see Chapter 3). Elsewhere, where the body is perpendicular to the ground (or within about 45° of perpendicular), the fabric tends not to be supported by the body; rather, it hangs against the body, suspended from a point of contact above. The location and number of points of contact can vary considerably depending on the angle of the body.[5]

In a weightless environment, gravity does not hold the garment in contact with the body at those locations; rather, the garment floats around the body, perhaps coming into contact with the body at certain locations, particularly if the fabric is elastic, but generally surrounding the body without being directly supported by it. As a result, fabric does not drape, but rather billows outward. Figure 4.7

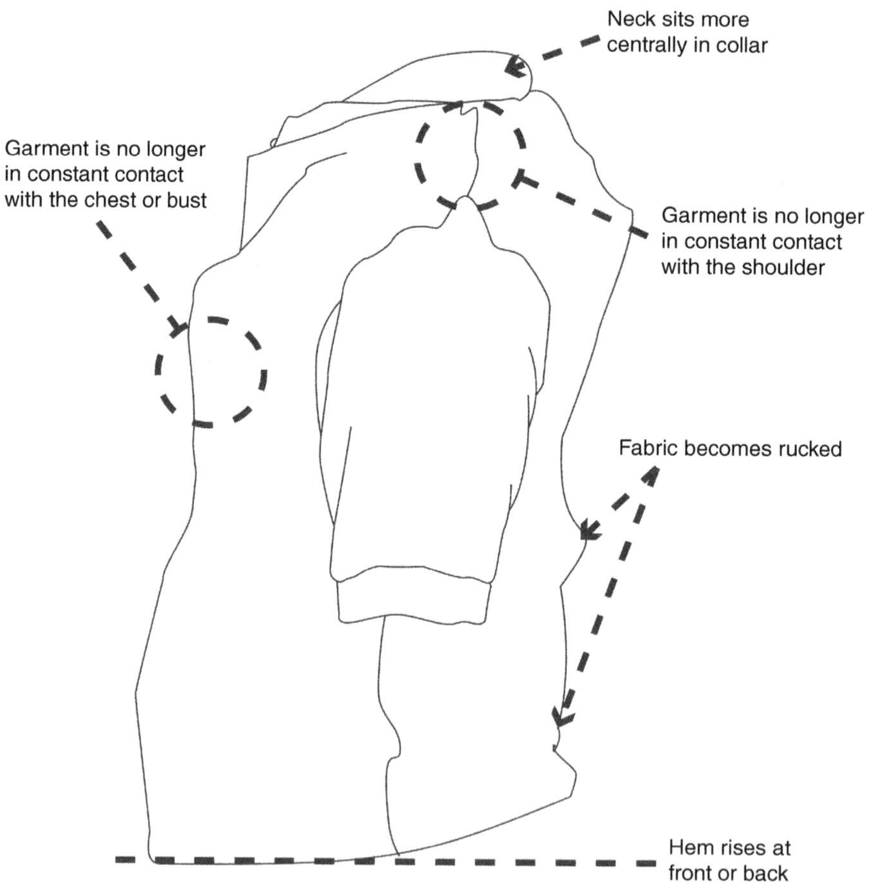

Figure 4.7 Diagram showing off-the-shelf polo short in microgravity, based on a survey of NASA videos and photographs of astronauts on board the ISS. In orbit on board the ISS, astronauts find that their off-the-shelf polo shirts contain but are not supported by their bodies. The garment is no longer weighted by gravity to the shoulder, and appears to float around the body, rarely skimming its contours

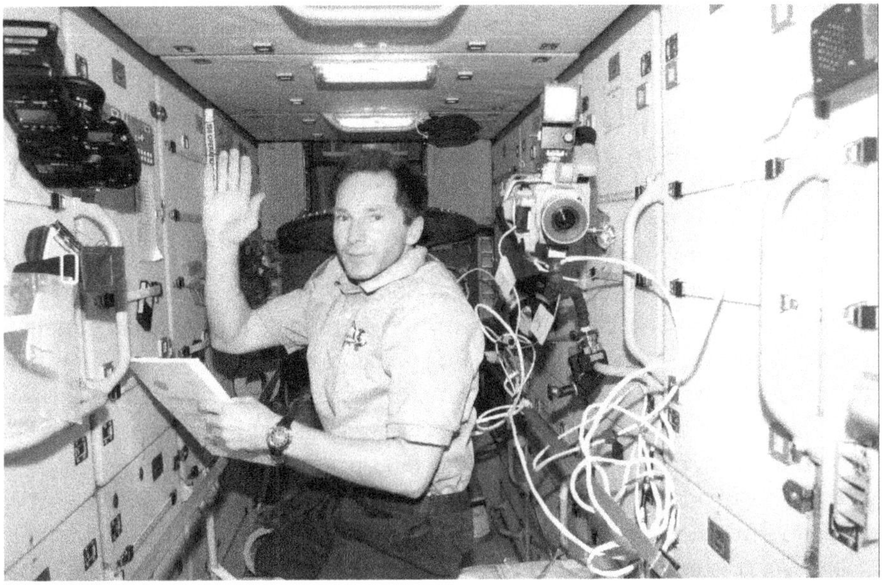

Figure 4.8 Cosmonaut Valery I. Tokarev at work in the Russian-built Zarya section of the ISS. His polo shirt balloons outward as if filled with air, surrounding his torso but not resting on it. His collar, which would, on Earth, rest on his chest, is instead lifted close to his chin. © NASA

shows how the same polo shirt might appear in microgravity. In this environment, the shoulder seam is no longer in contact with the shoulder, and the fabric dos not fall against the chest. Instead, it tends to balloon outward at these points. As the fabric does not drape downward, the hem is lifted upward, sometimes exposing the lower back (unless it is tucked into the waistband). The collar floats upward, so that it rests closer to the astronaut's chin and, due to the neutral posture, is more centrally positioned around the neck. These effects can be seen in more details in photographs of astronauts on board the ISS (see Figure 4.8). Astronauts' experiences tell us that as a result of not draping against the body, garments not only look different in space, but also feel different. Astronauts become so accustomed to living without the weight of clothes against their skin that some report physical discomfort from contact with everyday clothing after returning to Earth (Dunn, 2016; Harrington, 2016).

In the absence of adequate microgravity drape tests, digital technologies offer insights into the effects of different gravitational conditions on drape. Virtual three-dimensional garment simulation, that is already in use in the fashion industry, allows for a reduction in prototypes and samples, enabling more sustainable ways of working (Kuijpers and Gong, 2014, p. 1). Virtual dummies, with infinitely adjustable shape and height, are clothed with virtual garments, constructed from CAD patterns that may be tweaked, resized, and reshaped, in order to instantly visualize the effects of alterations within the digital environment, before manufacture of a real prototype. Users of CAD for garment design recognize that "accurate material representation is required when fitting a garment virtually," as the bending, shear, surface friction, and weight of a fabric will significantly affect the drape of a garment (Kuijpers and Gong, 2014, p. 18). Drape coefficients must first be measured, in a drape test, to ensure accurate representation of a fabric in the virtual environment. Kuijpers and Gong (2014, p. 22) identify three main factors that are required for the accurate virtual representation of a garment: accurate representation of the human body, accurate representation of the two-dimensional pattern from which the garment is virtually constructed, and accurate virtual representation of the fabric. "Seamless interaction between [these three] key elements" is required in order to realistically simulate a garment. This kind of digital simulation, by default, presents drape behavior as in an Earth-gravity environment. In order to enable simulation of a garment in different gravitational conditions, a fourth factor would need to be introduced, that is, the accurate simulation of reduced gravity.

In order to find digital methods of simulating the behavior of textiles in other gravitational conditions, one must look at activities that take place outside of the fashion industry, in the field of digital animation for games and film. Digital animators are required to design clothed characters for simulated on-screen environments, and objects within these environments are required to behave according to the particular physics of each virtual world, as determined by a "physics engine." For example, when two animated objects come into contact, they are designed to collide and rebound as if they were real objects. Digital animators have demonstrated concern for the accurate representation of fashion. Kuijpers and Gong (2014, pp. 8–9) cite the example of a spread in *Arena Homme+ Magazine*, in which a character from the computer game franchise *Final Fantasy* is shown clothed in a Prada S/S 2012 shirt that drapes photorealistically onto his virtual body. One common three-dimensional modeling software, 3DS Max, allows users to allocate a "cloth" identity to objects, and has presets for a range of textiles, including cotton, satin, silk, cashmere. It further allows users to manually adjust values such as stretch, density, or shear, so that they may tweak the appearance and behavior of a draped cloth. Once the properties of the cloth object have been set, the user may simulate dropping the cloth above a "collision object," so that it drapes over the object, much as it would in a drape test.

Despite the existence of these tools, the digital environment is not yet fully prepared to accurately represent the behavior of draped fabric in microgravity. In order to fully realize the potential of virtual modeling for the simulation of weightlessness for fashion design, expertise from two different fields—fashion and digital animation—must be shared. While digital animators are concerned with the realism of the bodies and garments that they generate for their virtual worlds, and have found effective methods of simulating weightlessness, they do not have the detailed knowledge of fabric and garment construction that exists within the fashion industry. Conversely, the fashion industry makes use of a range of methods of measuring the weight and drape of fabric that allow very specific differentiation between the qualities and potential behaviors of different textiles, but has not yet applied this knowledge in simulations of different gravitational environments. The fashion industry recognizes that a wide range of factors affect drape, not least, tensile and shear properties (Kuijpers and Gong, 2014, p. 3). Many of these properties can be measured, in part through drape tests, and so there is scope to make this data available in future physics engines for more nuanced simulations of the effects of weightlessness on garments than those that are in use in digital animation.

When digital simulations become more effective at accurately replicating the behavior of fabric in a range of gravitational conditions, they may become valuable in illustrating not only the forms that garments take in microgravity, but also during transition between different gravities. Ondřej Doule (2014a, p. 90), founder of space architecture magazine *The Orbit*, describes design for space as "defined by variable gravity." This chapter has so far addressed the microgravity conditions that exist on board the ISS, and that long-duration space tourists would experience on visits to orbiting vessels. However, it is important to acknowledge that many spaceflight participants will experience only short-term weightlessness. Passengers on board parabolic flights experience normal Earth gravity, interspersed with short periods of up to 1.8g (hypergravity) as the aircraft climbs, and shifting to zero-g in freefall, experienced as weightlessness (Karmali and Shelhamer, 2008). Commercial operators, such as Florida-based Incredible Adventures (2016), also offer to replicate the gravitational conditions of the moon (1/3rd Earth gravity) and Mars (1/6th Earth gravity) with shallower maneuvers. Passengers of future suborbital flights may experience approximately five minutes of weightlessness on board Virgin Galactic's SpaceFlightTwo, but will otherwise be strapped to their seats in familiar gravitational conditions (Howell, 2016). These passengers will dress and undress in Earth's gravity, and so their garments need to be suitable for Earth-gravity activities, including walking to the spacecraft, as well as for activities in a range of other gravitational conditions. On board parabolic and suborbital flights, when passengers are free to remain unseated during varying gravity, the body shifts from a horizontal lying position (or sometimes, an upright standing position) to the neutral posture, and so experiences in-between phases as they gradually become weightless.

A designer's awareness of gravity must, therefore, extend beyond a binary understanding of the differences between Earth-gravity and microgravity. Designers for space must be aware not only of the effects of microgravity, but also the many different gravitational conditions that may exist, both in spaceflight and, ultimately, on the surface of other planets or in artificial gravity. Moreover, they must accommodate the possibility that those different gravitational conditions will be experienced as part of the same journey, as in the "rapidly changing gravity levels in spaceship interiors for journeys from the Earth's surface into orbit" that "span from hypergravity to microgravity" (p. 93). Fabric will, in these varying conditions, sometimes drape from points of support on the body and, at other times, float free. It will be necessary for spacewear designers to design garments that are aesthetically and functionally suited to varying gravity. Garments must accommodate a range of postures, ranging from upright to seated to neutral, and must have a form that is suited to resting on, or being suspended around, the wearer's body. Some designers might approach this with compromise, limiting their engagement with weightlessness to forms that have a track record of use on Earth, or by making only small adjustments to Earthwear designs to accommodate the needs of a weightless wearer. Indeed, this is the approach that has been taken by existing parabolic flight operators, whose all-in-one flight suits resemble those already worn by pilots within Earth's atmosphere. Designers who are more willing to experiment might view this as an opportunity to design two garments in one—a single design that adopts different silhouettes as its wearer's environment changes. The potential for fluctuating silhouettes, that visibly evidence shifting gravitational conditions, poses a unique design challenge.

CONCLUSION

Designing post-gravity fashion

This book has identified the effects of microgravity on clothes, with a view to identifying the ways in which the fashion industry must reconsider its approach to design if it is to succeed in clothing the next generation of space tourists. Future space tourists will desire to be clothed in garments specifically designed for the extraterrestrial environment, and those garments will need to have fashion value. A new field of fashion, spacewear, will therefore emerge to cater to the needs of an audience that seeks to dress for the weightless environment. This new field will be populated by designers with specialist knowledge, including that contained within this book, and will be able to apply that knowledge in the development of designs for microgravity (and variable gravity) environments.

In just over half a century of manned space travel, designers and engineers have learnt enough to make educated predictions about the behaviors of a range of substances and artifacts in microgravity. Knowledge gained through the direct experiences of astronauts, and observations made via recordings, have contributed to an understanding of the effects of microgravity that is now deeply ingrained in scientific enquiry and in popular culture, and these are now familiar to television and cinema audiences, as well as those who have encountered the many images and recordings broadcast to a worldwide audience from the International Space Station (ISS) to the World Wide Web. This widespread familiarity with how objects behave in space contributes to expectations about commercial spaceflight, and what it will be like to experience microgravity first-hand. The chance to experience microgravity is one of the main motivating factors behind a desire for space tourism (Peeters, 2010, p. 1627). Early spacewear designers will have the task of designing in observance of formal knowledge of the effects of microgravity and in response to the informal expectations of audiences who have observed, second-hand, the effects of microgravity but, in many cases, have not yet experienced it for themselves.

With the knowledge that is currently available, this book has identified a range of key considerations that may define the emerging practice of spacewear design. These are considerations that are not present in Earthwear design, and so have not yet been characteristic of any particular Earth fashion trend. Some have, as this book has shown, been considered in other contexts, unrelated to space travel. For example, as was seen in Chapter 2, some designers have begun to consider the aerial view of their garments, and in Chapter 4, a seated position similar to neutral body posture is already accommodated in fashion for wheelchair users. Such examples demonstrate that these considerations are not alien to fashion designers, but they have not yet been explored in the context of design for space. The first generations of spacewear designers will need to consider, in combination, the way the body changes in microgravity, the absence of a downward gravitational force on fabric, and the full range of possible orientations of the body, along with the constraints imposed by the cost of transporting heavy or numerous garments to space, and safety concerns arising from the particular conditions of the journey. In addition, spacewear designs may also seek to incorporate new wearable technologies.

In writing this book, I have consciously avoided discussion of the wearable technologies, feeling that the distraction of wearable technology can compromise the handling of issues specifically related to weightlessness. Wearable technology, though still an emerging field, is the subject of more extensive existing discussion than fashion design for weightlessness, to the extent that, this author argues, a focus on technology has delayed engagement with issues arising from weightlessness. It is important, however, for future spacewear designers to be familiar with the available technologies that may complement or enhance the designs that are informed by this book.[1] Designers might find inspiration in, for example, color-changing fabric that could be used to indicate changes in surrounding atmosphere, including "heat or ion streams," or magnetic forces (Timmins, 2010, p. 200), or conductive fibers that can be woven into flexible material for wearables that record or transmit data (Stoppa and Chiolerio, 2014).

Many such proposed or existing wearable technologies monitor and respond to conditions, other than weightlessness, that exist in space (Arthur, 2016). A particular focus of wearable technology engineering has been long-term and long-distance space travel and, in particular, technologies for trips to the surfaces of other planets. Visitors to Mars might have their physical health monitored via Hexoskin's biometric shirts that monitor blood pressure, breathing rate, and heart rate, among other metrics (Howell, 2015). Mark Timmins (in conversation with the author) imagines emailing a hug to a relative who has settled on Mars, by transmitting a long-distance message to a mechanism in a garment that squeezes the wearer's body. Permanently separated from their friends and relatives, a Hug Shirt (already marketed by CuteCircuit) may be "the closest thing that they're going to get to connection with the people on Earth."

One technology that has been embraced by fashion designers in recent years is 3D printing. Given the presence of a 3D printer on the ISS, and the possible presence of zero-g rapid prototyping facilities on future commercial orbiting stations, future space tourists might not need to carry preformed garments on their journey from Earth (Prater et al., 2016). NASA (2014c) has already begun to print tools in the ISS, transmitting the design file from Earth to the onboard 3D printer. It seems likely that long-duration space travelers will eventually have the opportunity to download new designs for their clothing too. This may be the most cost-effective way of updating a wardrobe during a stay in space, given that the materials used to print a garment may be reused. 3D printing using materials that can be shredded, dissolved, and used to restock the printer would enable space travelers to recycle the garment that they wore the day before (Lipson and Kurman, 2013, p. 278). In this way, space travelers may have the ultimate flexible wardrobe, being free to restyle every garment they wear by reprinting it in a new design.

Wearable technology will augment the human body to improve the odds of survival in the new environments that we encounter as we begin to colonize space. Influential thinkers including Carl Sagan (1994), Steven Hawking (2016), and Elon Musk (2016) all tell us that "the attempt to settle space is an inevitability" (Smith, 2014), but having evolved to suit the Earth's environment, the spacefaring human will need to adapt, significantly and quickly, in order to successfully colonize other planets. Wearable technology will play a crucial role in that adaptation. Wearable robotic augmentations and exoskeletons might be worn to enhance human abilities (Salmoiraghi and Akin, 2012). FFD's Ted Southern strongly believes that wearable technologies should not only allow the survival of the human body in extraterrestrial environments, but should also enable us to perform operations of which we would not otherwise be capable (Kramer, 2014b). These wearables will be the interface through which we experience extraterrestrial environments, so that they do not merely assist human habitation of other planets, but also alter the ways in which the human body functions. Wherever atmosphere is not breathable, sights and sounds are filtered through the helmet and its components, creating the potential for significant enhancement or adjustment of the ways that humans sense their surroundings. Tim Leighton, Professor of Ultrasonics and Underwater Acoustics at the University of Southampton, observes that in the thin Martian atmosphere, in order for astronauts or colonists to be able to hear the sounds of potential hazards such as rock falls, microphones would need to be relocated. "You'd put the ears on the boots so that the sound that travels through the ground would be detectable," he tells the BBC (2017), illustrating the potential for spacesuit design to reconfigure human functionality.

In the decades that have passed between the Space Age and the new commercial space race, there have been a number of significant changes to our relationships with technology, and our understanding of the meaning and purpose of the body. The virtual body has emerged as a substitute for the real body. As N. Katherine Hales posited at the turn of the twenty-first century, we have become "posthuman" (Hayles, 1999). Gamers and social media users are accustomed to being represented as avatars—virtual surrogates that separate our intellectual experiences from our physical presence. In this new world, "bodies are obsolete" (Shaw, 2008, p. 87, after Stelarc and Stiegler). The mind surfs the digital realm while the body "remains docked, immobile at the interface" (Murray and Sixsmith, 1999, p. 318). Carter (2017) suggests that we may be less concerned about fashioning our own bodies, and more focused on styling our avatars. In a time when our lives are being increasingly lived in the virtual world, and that virtual world is becoming increasingly tangible (via the internet of things), many of our activities may be experienced vicariously through a virtual or physical avatar. As has already been demonstrated in social media and MMOGs (massively multiplayer online games), considerable investment is made into ensuring that avatars appear fashionable. These authors concern themselves, rightly, with the ways in which new technology, along with an increasing concern for sustainability, will shape future fashion. These developments and concerns will likely influence all fields of fashion, including spacewear.[2]

In our increasingly virtual world, space travel will be one of the few new physical experiences available to the masses, and spacesuits will be set apart by their physical interaction with the body. Spacesuits are the antidote to fears about the loss of the corporeal body in the internet age. Spacesuit engineering forces consideration of the materiality of the human body, and its most basic, physical requirements. The new frontier will force a return to material concerns. Physical survival will be the primary concern for crew traveling the vast distances to alien planets, and for the colonists that follow. In the early stages of colonization and terraforming, clothing will be as essential for our survival as food, water, and oxygen. It will ground us in the physical realm.

It may eventually become possible to consider our species not only as post-human, but also post-gravity. The experience of weightlessness is characterized by ungroundedness. Those who have experienced weightlessness on board parabolic flights have reflected that the sensation of being rooted to the ground is an integral part of the human experience. Kodwo Eshun (2005, pp. 28–29) observes that, since the dawn of mankind, gravity has "located the human species in a condition of subjective verticality." Taken for granted, groundedness has been part of the human condition. By exchanging Earth's gravity for the weightlessness of space, we change human nature. For Eshun, post-humanism can be achieved through the experience of weightlessness, that enables one to "re-experience gravity" as a "peculiarity."

Perhaps the most interesting wearable technologies are those that enable wearers to adjust their relationship with gravity, either by imposing the sensation of gravity on the body, in reduced gravitational conditions, or by freeing the body to behave as it would without gravity, where gravity prevents us from performing as we would without it. Man's dream of flight has driven the invention of various forms of air transport and, at a smaller scale, wearable technology such as jetpacks that allow the individual wearer to escape Earth's surface. Having achieved their goal of weightlessness, the spacefaring community has begun to recognize the value of being limited by gravitational forces.

The post-gravity human uses wearable technology to isolate its wearer from the effects of gravity, or its absence, to transcend concerns about those effects. The V2 Variable Vector Countermeasure Suit described in Chapter 4 can stabilize the body and apply resistance to create an artificial sensation of gravity, prompting a subjective orientation that may be counter to the equally subjective orientation of other bodies in the same space. Such post-gravity sensations can also be generated on Earth's surface, where wearable technology including the V2 suit has applications in stabilizing unsteady bodies. The desire to escape or to cling to the surface of the Earth, in ways that gravity would not normally allow, has yielded novelty inventions including magnetic hover boots, and Michael Jackson's (1992) patented "method and means for creating anti-gravity illusion," shoes that pin the performer to the stage, which enabled the musician and his backing dancers to perform an antigravity lean in the music video *Smooth Criminal* (1987). Jackson's invention bears resemblance to the shoe system that helped astronauts stabilize their bodies on board the Skylab space station (see Chapter 4).

Novelty technologies such as this are perhaps more commonly encountered within the realm of science fiction. The perceived relationship between groundedness and shoes has made them the focus of an array of science-fiction narratives. Fictional footwear technologies sometimes enable wearers to defy gravity, and at other times counteract the effects of weightlessness by grounding wearers to a surface. In *2001: A Space Odyssey* (dir. Stanley Kubrick, 1968, and in the same year, the novel by Arthur C. Clarke), Pan Am stewardesses wear "grip shoes" to keep them "down to earth" (Lathers, 2012, p. 159). Kubrick and Clarke imagine the interior cabin of a space shuttle lined with Velcro, and shoes soled with the same, so that stewardesses may tread walkways that run 360 degrees around the cabin's interior surfaces. Clarke (1968, p. 50) describes the "slight buoyancy" with which a stewardess walks, as she attaches and detaches her feet from the walkway. In Kubrick's film, the stewardess is seen to plant each foot firmly on the walkway, testing that it has reliably adhered to the surface before lifting and replanting the other. Much like the "hop" of Apollo astronauts' moonwalk (see Chapter 4), the character of her walk is determined by what she wears, which is in turn shaped by the peculiar nature of her extraterrestrial environment.

References to the fall of Icarus, both explicit and implicit, are commonplace in tales of hoverboots and other gravity defying wearables. Gravity shoes grant access to another world in the science-fiction film *Upside Down* (dir. Juan Solanas, 2012), but not without great risk to the wearer. The film is set in two neighboring worlds, each with its own gravitational field. Despite the proximity of the two worlds (there is a TransWorld skyscraper that bridges the divide), all matter is affected only by the gravity of its native planet, making it possible, through interaction with objects originating from a twin planet, to "fall up" or to "rise down." The peculiar physics of these two planets allows for the mid-air suspension of objects, weighted equally toward both worlds, and for the uncanny upside-down placement of objects. These man-made interventions in the two opposing gravities leads to a heat reaction, which ultimately causes objects to burst into flames if they remain in contact with objects from the twin planet. In order to escape his dystopian home world, Adam (Jim Sturges) must don shoes and a vest lined with weights that originate from his planet's twin. After spending too much time on the surface of the other world, Adam's shoes begin to overheat, the rubber soles start to melt, and ultimately catch fire. Adam must dive into the sea, where he removed the shoes, and is pulled skyward, back toward the surface of his own planet. Having defied the natural law of his fictional universe, Adam, like Icarus, plunges back to where he belongs.

Parallels with the Icarus myth have also been embraced by the real-life space-faring community. Mythological narrative has long been used to aid understanding of the human condition, and when Mankind extends his reach to the stars, Icarus provides an apt metaphor (Arnould, 2011, p. 33). Aerospace engineers fight a constant battle with fire, employing thermal protective layers to prevent shuttles carrying astronauts from burning up on reentry into Earth's atmosphere (which was the unfortunate fate of the Shuttle Columbia). References to Icarus remind the spacefaring community not to overstretch. They are "a warning against human pride," and, in light of plans to colonize Mars, a warning against anthropocentrism. Taking a more tempered view, Jacques Arnould proposes that it could also be "a simple call to good sense and caution, to preclude the dangers that go hand in hand with being care-free and young" (p. 34). The commercial space industry is still in its infancy, and on the occasions when it has overstretched its capabilities, the financial and human cost has been devastating.

It is important, however, not to associate all post-gravity or microgravity design with engineering or technology. Weightlessness, and hence spacewear, can be considered independently of wearable technology. While technology has enabled space travel, weightlessness is not itself a consequence of technology, and spacewear does not need to be technologically enabled in order to be made suitable for the microgravity environment. The zero-gravity coffee cup (see Chapter 1) is a fine example of a product intended for microgravity that makes no more use of technology than the cups we use in our own Earth-based kitchens. Clothing is among the least technologically enabled equipment on board the ISS (being, in the case of routine wear, off the shelf). Spacewear can be constructed using long-established dressmaking techniques. Indeed, we can identify many of the practices that need to be at the heart of spacewear design by looking at well-established or even historical concerns (see, for example, the tailoring techniques for stooped posture described in Chapter 4). Concerns that have been at the heart of all fashion design on Earth, including posture, body shape, and the ways in which the garment is arranged around the body, should also be at the heart of spacewear design.

The concerns that have been identified in this book with set spacewear design, apart from other future fashion practices, are summarized in the list below. This list builds on existing observations by researchers such as Annalisa Dominoni (2005, 2015, 2016), Timmins (2010), Orndoff (2013, 2015a, 2016) and Howell (2011, 2017), with the addition of concerns that have been examined for the first time in this book. Most of these are not yet under consideration by spacesuit engineers, but will need to be addressed when designing for space tourists, and perhaps other kinds of future spaceflight participants. This list of concerns, derived from the contents of the previous pages, can be presented as a set of guidelines for spacewear designers as follows.

1. Neutral body posture

Design for the body in neutral posture, much like the pose adopted by a snowboarder, with curved spine and bent legs (Dominoni, 2003, p. 279). Incorporate additional length on the spine and back of the crotch, outer elbows, and knees, and shorten the length of the front of the crotch, inner elbow, and back of the knees. Move the neck and armscyes forward.

2. Changes in body shape

Accommodate redistribution of volume around the body of the wearer. After long periods in microgravity, the body gains volume at the top through fluid redistribution, and loses volume in the legs through muscle atrophy. Measurements made on Earth should take account of these inevitable changes and, if appropriate, potential for adjustment should be designed into the garment, in the form of elasticity, adjustable fastenings, or drawstrings.

3. Alternative orientations

Consider the aerial view as equally important to the elevation view, as well as all other angles in between, including the view from beneath the feet. Bodies floating in a weightless environment will not have their feet planted on the ground, and may float toward each other from any angle. The elevation silhouette, as viewed from front or side, is as important as the silhouette viewed from any other angle.

4. Length as a radial measurement

Consider length not in relation to the height of the body but in relation to the garment's potential to extend outward in all directions. Untethered or nonelastic fabric will float free of the body.

5. Artificial contact points

Under Earth conditions, contact points are naturally created when the garment rests on the wearer's body. Without gravity to pull the garment toward the body's horizontal surfaces, garments surround the body without resting on it. If designers want a garment to remain in contact with the body at any location, they must create artificial points of contact at which the garment is drawn toward the body through elasticity, tethering, or tight-form. Similar points of contact are used in Earth fashions, as in waistbands, but may be needed in different locations when the garment is not also restrained by the downward pull of gravity. These contact points may need to be located differently to those in Earth fashion, to take account of (3) and (4) above. For example, to preserve modesty, a skirt may need to be drawn inward at the hem, or below the crotch.

6. Donning and doffing (dressing and undressing)

If designing for long-duration space travel, designers should consider how the garment may be donned and removed in the microgravity environment, perhaps taking cues from the modular design of the spacesuit. They must ask whether the garment can be removed independently by the wearer, or whether help is needed from a fellow passenger, or in the form of aids that might be designed in conjunction with the garment to enable independent donning (such as methods of stabilizing or tethering the garment to a stable surface during donning and/or doffing, incorporating rigidity to aid control, and/or including detachable sections so that the garment can be partially disassembled). Designers should also consider that, as the process of dressing and undressing does not need to be one of primarily vertical movement, entry and exit openings may be located anywhere on the body.

7. **Stabilization**

   Spacewear users will not always want to float freely around the cabin. They will need to stabilize their bodies to perform some activities. As the body cannot naturally adopt a standing position, stability will not normally be achieved via the soles of the feet, but by tethering any part of the body to a stable surface. Spacewear design must accommodate the need to tether or restrain the body. This can be seen as a freedom rather than a design constraint, as the point at which the body is stabilized can be located at any part of the body, and may therefore be positioned so as to minimize interference with the preferred silhouette. This fact may be of particular value to footwear designers, as it eliminates the need for a stable sole.

All of the above relate more to the aesthetics of form than to the technical or safety aspects of designing clothing for space. As well as the above considerations of form, spacewear designers will need to learn from existing advice on choice of textile technologies. The technical and safety requirements of wearables for IVA and EVA in space have been explored at length elsewhere, and designers aiming to design for space should consider the content of this book in conjunction with those other studies. There are a number of texts that consider textiles for space (Aspelund, 2017; Schneiderman and Griffith Winton, 2016; Tortora, 2015). There are also numerous studies by NASA and other space agencies devoted to the technical requirements of spacesuits and clothing for EVAs (Tortora, 2015). I have referenced only a few of these many studies here. Spacesuit design has been the practice of engineers rather than fashion designers, and hence most of these existing studies are written with an engineering audience in mind. It is the role of the next generation of fashion designers to understand the technical requirements outlined in engineering texts, and to negotiate a compromise between the fashion desires of space tourists and the safety demands of the spaceflight provider. Although their goals may diverge from those of NASA textiles specialists and spacesuit engineers, spacewear designers will still be responsible for following established guidelines relating to, for example, choice of textiles for spaceflight. There is considerable existing research on the suitability of textiles for space (Orndoff et al., 2014; Fratto, 2005), which has tended to find activewear textiles most appropriate due to their antibacterial qualities and high absorption capacity. Textiles should ideally be selected that follow NASA (2014a) flight safety regulations. Primarily, spacewear designers should consider the following features of textiles:

8. **Lint**

   Weightlessness can cause lint to float into cooling fans or other machinery and therefore poses a threat to safety. Low-linting fabrics should be selected, and should be shaven and laundered before use.

9. Flammability

The oxygen-rich environment, coupled with the dangers of evacuating a spacecraft mid-flight, make fire a more dangerous hazard that it can be in many Earth environments. Low-flammable materials should be selected.

10. Mass

The cost of transporting cargo, including clothing, out of Earth's atmosphere is likely to cause spaceflight participants to favor lightweight garments. Designers should consider the weight and mass of garments, as well as other means of reducing the overall mass of a wardrobe by, for example, offering adaptable clothing. Spaceflight participants may favor flexible and/or adjustable garments so that a smaller number of garments is required for a long-duration journey.

Some of these ten considerations are already taken into account in the selection of off-the-shelf clothing for astronauts, and in spacesuit design. Others have not yet been fully explored because they are less relevant to clothing for astronauts than they will be in fashion for space tourists. Concerns about the silhouette when viewed from different orientations, or for length as a radial measurement, arise largely when there is a possibility for free-flowing fabric. So long as spaceflight participants continue to wear relatively form-fitting clothing, such as flight suits, spacewear designers have no motivation to fully explore these fashion possibilities. It is only when designers are prompted to consider a variety of expressive forms, and a desire to visibly evidence weightlessness, that these observations become important, and will begin to alter designers' fundamental approaches to garment design.

In these future experiments, it will be essential not to draw a dividing line between the spacewear worn by astronauts and that worn by space tourists, or to assume that the list of considerations outlined above is applicable solely when designing for leisure travelers. As the space tourism industry expands, and long-duration travel becomes more common, the variety of motivations for space travel will also increase, as will the range of roles that spaceflight participants will play. There may be a range of additional space travel–related professions, other than that of an astronaut, and Earth professions will be adapted for extraterrestrial communities. Each of these professional communities will have different clothing requirements that differ from those of astronauts, and of space tourists (Timmins, 2010, p. 200). For example, one can compare the range of professions on board a cruise ship to those that might be required on a space-faring vessel of the future. There may be service staff, deckhands, salon operators, fitness trainers, administrators, sales representatives, hosts, and entertainers, in addition to the engineers and pilots whose role might be more comparable to those of current-day astronauts. The crew responsible for evening entertainment will have very different clothing demands than those responsible for preparing meals. They might, for example, require more elaborate garments that make theatrical use of weightlessness. As on Earth, each role will involve a range of different activities that require specific sets of clothing.

Equally, it is important to distinguish between the needs of different kinds of space tourists. Those who travel into space on a suborbital or point-to-point flights will only experience a short period of weightlessness. These passengers will likely dress on Earth before boarding the flight and will experience only a short period of weightlessness. For that reason, their garments will need to be designed for both Earth gravity and microgravity conditions. Passengers on long-duration space travel, or stays on board orbiting space hotels, will need to dress and undress in microgravity, and will experience constant weightlessness for long periods. Their garments will therefore not be required to function as such in Earth gravity, but will need to be donned and removed (or dismantled) without gravity, and preferably without assistance. The fashion needs of these two distinct tourist groups are therefore markedly different.

On board a vessel, where environmental conditions are artificially maintained, the development of spacewear will come hand in hand with the development of spacecraft interiors and other designed objects, so there is scope for collaboration between spacewear designers and designers from other disciplines such as interior architecture or product design. It is the spacewear designer's responsibility to be familiar with the features of the interiors in which their garments will be worn.[3] This has been true throughout the history of space travel, and will continue to be true as we develop a broader range of environments and facilities on board spacecraft and space stations. Developments in spacesuits have been prompted by improvements in the technologies and conditions of space travel, and by the increasing length of time that astronauts are now able to spend in space. Gravity Loading Countermeasure Skinsuits, for example, were developed in order to alleviate the adverse effects of long stays in microgravity, years after the ISS made such long stays possible (Waldie and Newman, 2011, p. 722). Need-driven design such as this is likely to take place in spacewear as long as the effects and possibilities of space travel are not wholly known. Speculative plans for microgravity sports facilities, such as those outlined by Collins et al. (1997) and the artificial gravity environments that might be possible on board a rotating wheel space station, such as the Gaia Class Space Station proposed by United Space Structures (O'Callaghan, 2015), all describe possible new environments and conditions that will prompt further developments in spacewear.

This book has acknowledged that the space tourism industry will, at least in its initial phase, cater to the luxury market. For this reason it may be helpful to look to other luxury markets to ascertain the needs and aspirations of the audience. The first generation of space tourists will be a privileged few, and will be able to take pride in the fact that they were among the first to extend human culture beyond our home planet, and may be able to affiliate themselves more with astronauts than with casual consumers of the space-travel experience. A decade ago, Linda Billings (2006, p. 164) predicted that space tourism could pose a threat to the noble ambition of space exploration, in the frontiersmen tradition, by positioning space travel as "joyrides for the ultra-rich," with no greater cultural or social value than as a "status symbol." At the time of writing we are not much closer to understanding whether these two visions of future space travel are incompatible. SpaceX's plans for the colonization of Mars seems to provide evidence that commercial spaceflight operators will contribute to the continuation of mankind's "heroic journeys of discovery." Elon Musk's (2016) plans have been driven by an evidently altruistic desire to "make life interplanetary." It may be, therefore, that mass space travel emerges simultaneously in a number of different forms, catering to professional, altruistic, and leisure travelers.

Until the cost of spaceflight is reduced, it will be impractical, if not impossible, to test most spacewear design outside of the Earth's atmosphere. Parabolic flights currently offer the closest available simulation of the environment on board a space station. On board a parabolic flight, a dressed body can be subjected to short periods of free fall. However, the current cost of parabolic flights may be prohibitive for many designers, or at least such that it must be reserved only for final testing or display of a finished garment, and not at intervals during development. This book has proposed a range of alternatives that provide adequate substitutes for space-based testing in the interim period (such as underwater testing), before a true microgravity environment is more readily available to designers and manufacturers. Spacewear designers of the future are likely to have access to more advanced testing facilities than those currently available to fashion designers, offering microgravity testing for longer periods than parabolic flights or drop towers, and at a fraction of the current cost. They may also have more advanced digital simulation software available for virtual testing of their designs, or more frequent and cost-effective access to orbital research facilities. For now, designers are only able to make informed guesses about how their garments will behave in microgravity. This is not dissimilar to how scientists and engineers have developed numerous other artifacts destined for space, relying on underwater testing, short-term testing on board parabolic flights or in drop towers, imperfect digital simulation, and the application of logic, informed by observation of how other objects have behaved in microgravity conditions.

The limits of our currently available testing environments, and the fact that so few tourists have yet made the journey to space, are reminders that mass space tourism is still science fiction. Any plans for the particulars of the space tourist experience are speculative. We are still today, as we were in the Space Age, "lying in the hammock and gazing at the moon . . . so close, yet so impossibly remote" (Wachhorst, 2000, p. 60). Early generations of space tourists will have grown up with science-fiction visions of space travel, and their journeys to space will be much like taking part in their own science-fiction movie. Those associations with science fiction may drive some early space tourists to seek clothing that emphasizes the fantastical nature of their voyage. The dreams of early space tourists will also be informed by established space travel imagery that presents the astronaut in his spacesuit as a heroic frontiersman. For them, a trip to space may be an opportunity to play at being astronauts. They may desire clothing that connotes heroism, in the spaceman tradition, or that hints at scientific purpose. There may be, therefore, a demand for a functional aesthetic, with similar references to the spacesuit as those that appeared in Space Age fashion, such as helmet-like caps.

With all of these things in mind, what might future spacewear look like? In her survey of space art, Annick Bureaud (2006) hypothesizes that, given the possibility of free movement in three dimensions, weightlessness may prompt a transition from planes to spheres as the basis of artistic practice. Silhouettes might, following Bureaud, become spherical, or more three-dimensional in other ways. With no ground on which to stand, wearers might choose garments that extend outward, past the soles of the feet. They may even move so far beyond the forms common to Earth fashion that the fashion industry will need to invent new language with which to describe these strange new silhouettes. These predictions are all, however, informed by the experiences of designers and artists who have made a temporary engagement with weightlessness from the perspective of someone whose life experience has been predominantly on Earth's surface. We are, Mark Timmins reminds me, "at the very beginning of a revolution."

This infant stage in the spacewear revolution will be characterized by the extent to which design and manufacture must take place on Earth. At present, the space industry is based on the Earth's surface, including design and manufacture. The presence of a 3D printer has led to the experimental production of tools on board the ISS, and this is the first small step toward extraterrestrial manufacturing. We are a long way from establishing an off-world fashion industry, in which clothes are designed and manufactured in microgravity conditions. That second phase of the spacewear revolution will introduce a new set of problems related to manufacture. Astronaut Karen Nyberg carried sewing supplies with her to the ISS, with the aim of sewing a small quilt block on board, and found the experience of sewing in weightlessness to be very different to the equivalent, familiar experience of sewing with gravity to hold tools and materials in place. “Now that I have tried my hand at sewing in space I can say one thing with certainty—it’s tricky!” She describes “challenges with cutting, piecing and stitching.” She experienced trying to cut a swatch of fabric to shape, when “you can’t lay things down and measure and cut” (NASA Johnson Space Center, 2013). Once a need for space-based garment manufacture emerges, so too will new dressmaking process and technologies.

Current fashion practice still views spacewear through the lens of science fiction, informed more by the aesthetics of spacesuits and fantasy than by investigation into how weightlessness affects clothing. Given the youth of the commercial space age, it is perhaps inevitable that design for space is predominantly fiction or prediction. The image of the human body in microgravity is still seen more often in science fiction than in reality. For most of us, our understanding of spacewear is informed less by the realities of space exploration than it is by our hopes, dreams, fantasies, and fears of the universe beyond Earth’s atmosphere. Speculative designs for spacewear, like science fiction, extend “knowledge of contemporary technology into an imagined future” (Gooden, 2012, p. 13). Even those spacesuits recently promoted by NASA, the Z-2, are intended for a mission to Mars which may never happen. Though commercial space travel is very real, it is in its infancy, and so design for space is often more informed by speculation than experience. This book aims to encourage preparedness for mass space tourism, and perhaps to influence future designers’ approaches to spacewear. These investigations are becoming increasingly pertinent. Now that we have begun to extend our territory into space, it is inevitable that weightlessness will become part of our future, including the future of fashion.

GLOSSARY

Astronaut Professional who serves as pilot or crew of space vessel. In most cases, the term excludes space tourists. In 2002, agencies involved in the ISS introduced the term "spaceflight participants" to distinguish space tourists from professional astronauts (Freeland, 2005, p. 11). Richard Garriot, who travelled to the ISS as a space tourist in 2008, has proposed the terms "private astronaut" and "civilian astronaut" to more closely align the activities of orbital space tourists with those of professional astronauts (Fisher, 2009, p. 60). The term is used in this book to include the Russian *Cosmonaut*, and Japanese *Taikonaut*.

Drape The way in which a fabric falls or droops "under the action of gravity" (Jiang, Cui, and Hu, 2012, p. 661). Drape is a consequence of the weight, thickness, shear, and surface friction of a fabric. Drape occurs when fabric hangs off an object, and is therefore partially supported by it, as when gravity forces a garment to hang from the body of its wearer.

Drop Tower A tall tower used to produce a short period of relative weightlessness through free fall by enabling a capsule to be dropped or catapulted inside a tube that runs the height of the tower. One example is the ZARM tower in Bremen, Germany, which is 146 m tall, delivering 4.74 seconds of near-weightlessness inside the capsule.

EMU Extravehicular Mobility Unit, more commonly referred to as a spacesuit. EMUs are modular systems worn for extravehicular activity (EVA), such as a spacewalk or activity on the surface of another planet, capable of supporting life and protecting the body from the dangers of space. They typically consist of a helmet and visor, upper torso, lower torso, gloves, and boots, and life-support system. The suit's many layers consist of liquid-cooling liners, pressure bladder, and outer thermal micrometeoroid layers that reflect heat and protect the body from micrometeoroids.

EVA Extravehicular Activity, also commonly known as *spacewalk,* but can include activities on the surface of other planets, as in a moonwalk. EVA can include time spent tethered to the exterior of the space station, attached to a robotic arm, or untethered in a Manned Maneuvering Unit. EVA typically serves the purpose of repairing or upgrading the space station.

Earthwear All clothing that is designed with the (generally implicit) purpose of being worn within the earth's atmosphere. The term employed in this text to differentiate clothes designed to be worn on Earth from those that are explicitly designed to be worn in space.

Flight suit A regulation one-piece worn during flight or spaceflight. Flight suits may be flame retardant but do not otherwise offer the protection that may be offered by other suits such as *pressure suits*.

Free fall The condition of falling as a result of being pulled by gravity, with no other force acting to direct the motion. Free fall can be experienced as weightlessness when it occurs in the absence of drag. For example, a passenger on board a falling airplane is contained within the cabin, and does not experience the drag of the air outside. She falls at the same speed as the cabin and is therefore not falling relative to the cabin but, rather, apparently weightless.

Gravity A force that attracts objects toward the center of a planet or other body. Due to their small mass relative to Earth, the gravitational forces on the moon and Mars are weaker than on our home planet. The further an object travels from Earth or any other body, the less it is affected by that gravity. It is not possible to entirely escape Earth's gravity, as the force decreases gradually at distances away from our planet. Traveling away from Earth, a vessel may consider itself to have escaped Earth's gravity as eventually, beyond approximately 2.2 million km distance, the gravity of the sun will dominate. A decrease in gravitational forces can be simulated in *free fall*. Increased gravitational forces can also be simulated mechanically, as in, for example, a centrifuge.

Gravity-loading The application of pressure and/or resistance to mimic the effects of gravity, usually on the body to avoid musculoskeletal degeneration, as in a Gravity Loading Countermeasure Suit.

ISS International Space Station. The ISS is the only man-made, habitable satellite in orbit around Earth. The ISS maintains a distance of between 200 and 250 miles above Earth's surface, and is in constant free fall, with the result that its inhabitants experience weightlessness. As it is of modular construction, the ISS has the potential to be expanded or upgraded, as when, for example, Biggelow Aerospace attached an inflatable pod to the Tranquility module in 2016, with the aim of expanding the ISS's habitat.

The ISS replaced previous space stations, Mir (1986–2001) and Skylab (1973–79).

IVA Intravehicular Activity, which is, at the time of writing, most likely to include activities carried out within the cabin of the ISS.

Microgravity Severely reduced gravity, resulting in the condition of weightlessness, often improperly described as "zero gravity." Since all objects with mass generate their own gravity (however small), passengers on board spacecraft would feel weightless, but would be subject to imperceptibly small gravitational forces, even when located a long distance from any planet.

NBL Neutral Buoyancy Lab. Underwater training facility used to prepare astronauts for *EVA*. Water does not provide a perfect simulation of a weightless environment, so divers are employed to counteract the effects of water drag and pressure, to more effectively imitate the neutral environment that astronauts encounter during *EVA*.

Orbit A state of permanent free fall, in which an object maintains a fixed distance from Earth, as a result of perpetually falling toward Earth, while the earth rotates away from it.

Parabolic flight Flights within Earth's atmosphere that follow parabolic arcs so that passengers may experience varying gravity within the cabin, including weightlessness when the aircraft enters *free fall*. Parabolic flights are currently the only way for a human to experience weightlessness without travelling into space, but only offer 20–30 seconds of weightlessness per parabola.

Pressure Suit A one-piece that protects the wearer from unexpected or planned depressurization by applying mechanical pressure or by filling with pressurized gas. Pressure suits filled with gas are bulky and allow limited mobility. Mechanical pressure suits such as Dava Newman's (2012) Biosuit are proposed as a form-fitting and flexible alternative for wearing in low-pressure atmospheres such as on the surface of Mars.

Point-to-point spaceflight Suborbital travel between Earth destinations via space, that would significantly reduce the length of some long-distance journeys.

Reduced-gravity aircraft See *parabolic flight.*

Suborbital spaceflight Flight that enters space but does not travel high enough to enter orbit.

Space Tourist A private citizen who has paid (or been funded by a third party) to travel to space, principally for leisure purposes. To date, there have been seven space tourists, beginning with Dennis Tito in 2001. SpaceX has announced plans to send a further two private citizens into space, on a circumlunar journey (Yuhas, 2017).

Spacesuit There is no universal spacesuit design, as each is designed for a specific environment and purpose. Typically, a spacesuit serves the role of a personal, miniature spacecraft, offering complete life-support and protection from the low- or zero-pressure environments, extremes of temperature, and radiation. See *EMU*.

Spacewalk Activities that take place outside of a spacecraft, usually as it is orbiting Earth. See *EVA*.

Spacewear Clothing that is designed with the explicit intention to be worn in space. This does not include *Earthwear* that has been intended for wearing on Earth but that has been later selected for its suitability for use in space.

Weightlessness The phenomenon of being free from noticeable effects of gravity, as when in *free fall* or when located at such a distance from a planet that its gravitational influence is negligible. It is not possible to simulate the experience of weightlessness on Earth's surface, but objects can enter *free fall* if dropped from a great height, as in a *drop tower* or on board a *parabolic flight*.

Zero-g Abbreviation of "zero-gravity." Sometimes abbreviated further to "0-G." The term implies an absence of gravity, but since a small amount of gravity exists throughout space, it has come to be used synonymously with *microgravity*.

Zero-g flight See *parabolic flight*.

NOTES

Introduction

1 The race to Mars will see commercial agencies competing with the government agencies of the United States, India, Russia, and China (Friedman, 2013). In 2015, NASA laid out plans for the possible future colonization of Mars, describing the project as "a natural evolution of prior decades of space exploration." The report declares that "Mars is an achievable goal" and is the first step toward "extend[ing] human presence deeper into the solar system" (NASA, 2015b, p. 1). The plan also stresses its reliance on commercial partners and outlines its ambition to "foster an emerging commercial space industry that will mature to support future pioneering missions" (p. 15). Commercial agencies are developing with similar goals in mind. Indeed, Elon Musk recalls that "the reason SpaceX was created was to accelerate development of rocket technology, all for the goal of establishing a self-sustaining, permanent base on Mars" (Howell, 2017a). His plans for the journey to Mars, outlined in September 2016, involve "a public private partnership" that may exploit his existing relationship with NASA to make Mars a feasible destination for both parties (Musk 2016).

2 The length of time spent weightless during a suborbital flight will vary depending on the altitude reached by the spacecraft. Craft flying to 110 km will offer three minutes of weightlessness; craft flying to 120 km offer four minutes of weightlessness, and so on (Crouch at al., 2009, p. 446).

3 The first space tourist was Japanese journalist Toyohiro Akiyama, who spent a week in orbit in 1990, and was followed eleven years later, in 2001, by Dennis Tito (Gibson, 2012, p. 59). However, the total number of space tourists is still small (seven between 1990 and 2009, and another two tentatively scheduled for SpaceX flights in 2018 or 2019).

Chapter 1

1 The term "clothes" is used here to include a spacesuit, as it is a worn object, but it will be important for the future fashion industry to distinguish spacewear designers from spacesuit engineers.

2 Despite the diminishing support for the space program, optimists continued to envision ongoing extraterrestrial ventures, predicting the kind of commercial engagement with space that now seems inevitable. Following the success of the Moon landing, President Richard Nixon envisioned a future of “routine space travel,” leading to and eventually becoming reliant upon “exploitation of . . . off-world environment[s]” (Woods, 2009, pp. 26–27). Nixon declared that a shuttle program would be NASA’s “next logical step” (p. 26). In 1972, anticipating a future in which interplanetary space travel would become the norm, he declared a multibillion dollar investment in a program that ultimately proved to be unrealistic in its goals of conducting seventy-five flights per year (p. 29). At the same time, both the United States and Soviet Union were engaged in a competition to establish a permanently habitable off-world base: an orbital space station. This would make it possible to study the long-term effects of weightlessness on the human body, and to conduct experiments on how other substances would behave in those same conditions. Space stations were to join the many satellites already in low-Earth orbit.

3 Another of Shaw’s observations is worth considering here. After the aging or death of the astronaut (perhaps more often in science fiction than in the short history of real space travel), the empty spacesuit evidences Man’s ability to create technology that will outlast the men who operate it. In this way, writes Shaw (2004, p. 126), the spacesuit represents “the triumph of technology over bodily disintegration.”

4 SpaceX’s Falcon 9 reusable rocket toppled over on landing and exploded in 2009, but SpaceX downplayed the incident as a learning experience (Kluger, 2009). They later succeeded in a successful Falcon 9 landing after delivering communications satellites to orbit in 2015, and again after launching the Dragon spacecraft to the ISS in April 2016 (SpaceX, 2016).

5 The Bigelow Expandable Activity Module (BEAM) will be carried to the ISS on board a SpaceX Dragon cargo ship.

6 The commercial space race has claimed several victims, including Rocketplane Limited, an Oklahoma-based company that declared an interest in “hyper-space couture,” but filed for bankruptcy in 2010. Rocketplane’s assets were purchased and the company was relaunched under the new name of Rocketplane Global.

7 Sources dispute whether *The Moon Museum* was actually left on the lunar surface.

8 Videos of projects funded by the Arts Catalyst are useful in understanding how clothing behaves in weightlessness. Jem Finer and Ansuman Biswas wear loose fitting clothing in *Zero Genie* (2001), and Morag Wightman wears a skirt in *Falling without Fear* (2001).

9 Translated from French: “*la fluidité extraordinaire des mouvements*.”

10 Translated from French: “*Dans un univers à trois dimensions, sans poids, il faut se cré er des ré fé rences é gocentré es subjectives puisqu’il n’y a plus de centre de gravité . On peut se repré senter une verticale subjective ou un point situé sur cette verticale. Il n’y a plus de ré fé rent universel, tout est relatif et chaque personne a la liberté de construire ses propres structures sur l’axe subjectif de la verticalité . À partir de cette construction interne, on peut appré hender l’espace externe et ré agir par rapport à lui.*”

11 The behavior of floating paint in microgravity was the subject of research by the first British-born space tourist Richard Garriott. On board the ISS, Garriott released droplets of paint into a cube lined with white paper, to observe how and where they would land. "As the paint dried it formed spheres rather than falling flat onto the paper" (Garriott, 2008).

12 Devices such as the ISSpresso machine, a zero-gravity espresso machine developed by Argotec (2014) and Lavazza, introduce the possibility of café culture in weightless environments. However, by dispensing coffee directly into a pouch, the ISSpresso deprives the drinker of the aroma that is so important in a coffee-drinking experience. Moreover, pouches do not allow for rituals that form part of the culture of drinking. Rituals such as toasting, which can be so important in establishing and maintaining social bonds, and during the celebration of uncommon achievements (so many of which occur on board the ISS) require drink to be contained in a rigid vessel.

13 "The number of hours taken to complete each garment in [Margiela's *Artisinal*] line is recorded on the label," so that "the lengthy procedures of working these cheap, mass-produced and disposable materials by hand into unique garments gives them rarity and value" (Groom, 2011, p. 505).

14 The problem can be observed in other examples of nonregulation garments worn on parabolic flights. YouTube footage of passengers wearing ball gowns on board Japanese parabolic flights illustrates how a loose skirt can lift, so that its hem pools around the wearer's neck and head, uncovering the lower parts of the body, and obstructing the wearer's vision. Loose-fitting shorts can present similar problems. Footage of tours of the ISS illustrate the potential for compromised modesty, as astronauts are filmed diving through the cabin head-first, exposing camera operators to the billowing hems of their shorts.

Chapter 2

1 Many issues of *Vogue* during this era seem to use the term "weightless" to refer to lack of visual weight, not just low density. The February 1960 issue of *Vogue* equates weightlessness to "bulklessness," in their description of a figure-hugging Adele Simpson dress (p. 83).

2 Through this subtle reference to the history of air travel, the balloons extend Chalayan's engagement with aviation, a theme that has underpinned many of his collections. His references to aircraft and aerodynamics feature most prominently in "Aeroplane Dress" (*Echoform*, Fall/Winter 1999).

3 While a rigged display does lift a garment off the ground, and off the human form, it can also reinforce the presence of gravity. At Silicon Valley Fashion Week, Betabrand sent its 2015 collection down the catwalk suspended from flying drones (Carson, 2015). As they fly, the drones' rotor blades blast air into the garments, causing them to inflate and undulate. Despite the use of gravity-defying drones, Betabrand's garments are clearly flying rather than floating, and subject to the pull of gravity. As they are lifted upward by drones, the garments hang limply downward. Further evidence of gravity's pull is seen in the effects of the rotor blasts, which illustrates the effort required to keep the drones aloft.

4 These experiences could not be replicated on dry land, where the suits are too heavy for a wearer to operate.

5 Such a wide circumference would be impractical on board the current ISS, or a parabolic flight, where cabins are tight and cramped. When commercial space travel becomes more common, more spacious interiors may be introduced that allow more room per passenger. Historically, the use of a wide circumference has been desirable in womenswear (often supported by a pannier or crinoline), and has signified class. A large skirt requires a large space in which to move—yet another sign of the luxurious lifestyle of the aristocracy. Notoriously, the pannier of the rococo era was sometimes so wide that women were unable to walk forward through doorways, leading to a requirement for more luxuriously proportioned doors and corridors (Mackenzie, 2009, p. 21). Since space tourism is emerging as a luxury market, the opportunity to occupy a large area on board a vessel, with body and clothes, might be an even greater signifier of wealth and exclusivity. A long hem, ballooning around the body, might emphasize the spaciousness of a luxury vessel, in contrast to the overcrowded cabins occupied by professional astronauts on board the ISS.

6 While there are few examples of footage of loose clothing worn on board the ISS, footage of loose hair can be revealing. Although fashion garments have not yet been worn in space, several astronauts with long hair have spent time on board the ISS, offering opportunities to predict how loose fabric might behave in the same environment. For an interview broadcast live from the ISS in 2011, astronaut and flight engineer Catherine Coleman wore her hair loose (CBS, 2011). Her shoulder-length hair fans upward and outward, forming an irregular sphere around her scalp. In contrast to the movement that can be in hair on Earth, in and out of water, Coleman's hair is surprisingly motionless. This gives her hair the appearance of being held in place by a strong hairspray. Although it does maintain a far more constant shape than hair underwater, Coleman's hair can be seen to shift slightly at the tips when she moves her head. It is also worth noting the effects of weightlessness on Coleman's T-shirt and jewelry in this same video. Her T-shirt is held against the body at the waist as it is tucked into her trousers, but elsewhere it balloons around her body, adding volume around her upper torso. Her long, rigid earrings, designed to hang downward, instead float almost horizontally outward, and her neck chain drifts languidly below her chin.

Chapter 3

1 Any future spacewear designer might want to consider how a modular approach could offer similar advantages for IVA spacewear. For future commercial spacewear, modular designs would offer a number of practical advantages, without necessarily compromising on style. Given the cost of transporting clothing into space, modular garments might reduce the cost of a passenger's luggage by reducing the overall load of his or her wardrobe.

2 Backers pledged a total of $27,632 to support the project, in exchange for rewards ranging from postcards and T-shirts to, in one case, a fully functioning suit. The rewards were of a lesser value than the donations that each supporter contributed, so that the profit could be used to fund the development of the new spacesuit.

3 The cost of producing a spacesuit is so high that it has determined who has been allowed to travel into space. Astronauts are required to be between 62 and 75 inches in height, so that they can fit into a modified spacesuit rather than requiring a new one to be constructed from scratch (Steimle and Norberg, 2013, p. 260).

4 Commercial spacesuits can be considered not just in relation to science fiction, but also in relation to other forms of costume spectacle. In the commercial sector, where spacesuits are engineered by independent businesses such as FFD, spacesuit designers come from an array of creative backgrounds. Artist and designer Ted Southern, cofounder of FFD, previously designed technical costumes for Cirque du Soleil and the National Opera. His work for theatre incorporates LEDs and inflatables rigged to transform costumes, or set them in motion, as a performer deploys a CO^2 canister that is concealed within the costume. He has necessarily been concerned with spectacle, developing wearable technologies to thrill and surprise his audiences. For Southern, theatre costumes and spacesuits exist on a continuum (Kramer, 2014b). Both must allow the wearer to move freely, and both involve, in Southern's experience, "bodycentric . . . assemblies" (Oberhaus, 2015a).

5 All COTS equipment must meet a list of safety requirements, including requirements related to flammability, sharpness of corners and edges, pinch points, and so on. In conversation with the author, Libby Jackson of the UK Space Agency describes the problems they encountered when trying to meet these requirements for objects sent to the ISS alongside British astronaut Tim Peake. The team was not permitted, for example, to include among the objects a scratch-and-sniff card, due to the fact that there would be no way of expelling the odors that the card would release into the cabin.

Chapter 4

1 There are exceptions to this rule. Shoes and boots for activities such as horse riding are designed with the knowledge that the rider's body will not always be weighted to a planar surface. Pointed shoes are intended for dancers whose point of contact with the ground will shift between sole to toe, and so have a second planar surface at the tips of the toes.

2 Or that looser fabric will hang from stiffer, structured parts of the garment.

3 The way in which the garment hangs is known as the "balance," and the way it molds against the body is known as the "set" (Stecker, 1996, p. 199).

4 Asymmetrical garments are cut so that they appear unbalanced, with one part of the hem or seam intentionally at different heights.

5 The number and location of points of support vary significantly depending on the pose and angle of the body or its limbs. Traditionally, garments are constructed so that the fabric plane is parallel to the force of gravity (Mei et al., 2015, p. 2). In Earth gravity, fabric drapes in the direction of the floor, regardless of the angle of the body. A garment would, therefore, fit a body differently if the wearer were to hang upside down (see, for example, how garments hang from the body in vertical fashion shows, Chapter 2). When the body is in motion, additional points of contact are created, where gravity causes other parts of a garment to rest upon other body parts. For example, when a knee is raised for walking, the draped part of a skirt that might otherwise hang loose comes into contact with the knee and/or thigh, and is lifted so that the fabric now hangs from the leg as well as from the waist. Depending on the speed and character of the body's motion, the garment's behavior around the body can appear to duplicate the behaviors exhibited in microgravity.

Conclusion

1 I can recommend, for example, the projects collected in Sabine Seymour's (2008) *Fashionable Technology*, or the predictions made in Bradley Quinn's (2012) *Fashion Futures*.

2 The technology that enables space travel will one day, in the most literal sense, make us post-human. The shape of the human body, and the character of the actions that it performs, are all products of gravity. Long-duration weightlessness weakens the muscles and bones in such a way that it becomes difficult to handle Earth's gravity on return, and it is months before astronauts can once again function normally in their native Earth environment. Habitual activities including walking and talking must be relearned as the returned astronaut undergoes daily medical tests and physical rehabilitation (Howell, 2013). Those colonists who may soon settle on Mars will find that the effects of weak Martian gravity will make the return to Earth even more difficult, if not impossible (Holick, 2000, p. 1570). Colonists' offspring—the first generation who will be able to legitimately call themselves Martians—will have skeletal structures developed for a gravity that is approximately a third of Earth gravity (Impey, 2015). From the moment of the first extraterrestrial birth (whether it be on Mars, in orbit, or any other location with non-Earth gravity), a new subspecies will begin to emerge that diverges from the evolution of humans on Earth (Solomon, 2016). If humans do succeed in becoming a multi-planet species, the different gravitational conditions of each planet will mean the existence of a variety of body shapes and sizes, necessitating the need for design that is tailored toward the particular body type of each subspecies.

3 Historically, there have been developments in fashion that have prompted developments in design in other fields, and vice versa. A tongue-in-cheek article in *Munsey's Magazine* (Forrester, 1908, p. 524) lamented the narrowing of skirts in response to the narrowness of doorways in domestic interiors, and proposed that architects should revert to planning wider doorways for the sake of those wearing not only wide skirts, but also wide-brimmed hats.

BIBLIOGRAPHY

Abiera, M. (2016), "Zero-G Clothes," email correspondence between Barbara Brownie and Mara Abiera, Operations Coordinator for ZERO-G, April 8–11.

Adams, C., and R. Jones (2014), "From International Style to the International Space Station," *Architectural Design*, vol. 84, no. 6, pp. 71–74.

AFMC (Air Force Materials Command) (1993), "COTS: Commercial Off-the-Shelf Guide," NASA Electronic Parts and Packaging (NEPP) Program Website. http://nepp.nasa.gov/docuploads/1219C61B-7337-48C4-8760E6456F861839/COTS%20guide.pdf (accessed August 18, 2016).

Aldrich, W. (2015), *Metric Pattern Cutting for Women's Wear*, 6th edn, Chichester: Wiley.

Alvarez, E. (2016), "Virgin Galactic Taps Fashion Brand Y-3 for Space Suit Designs," *Engadget*, January 14. https://www.engadget.com/2016/01/14/virgin-galactic-y-3-space-suits/ (accessed October 7, 2016).

Ara, P. (2016), "Space Textiles, Space Patterns," *Couture in Orbit*, European Space Agency. http://www.esa.int/Our_Activities/Human_Spaceflight/Couture_in_orbit/Space_textiles_space_patterns (accessed August 9, 2016).

Argotec (2014), "Coffee in Space: The Italian Espresso in Orbit with Argotec, Lavazza and the Italian Space Agency," *Lavazza.com*. http://www.lavazza.com/en/.content/document/pdf/PR_Italian-espresso-in-orbit.pdf (accessed July 6, 2016).

Arnould, J. (2011), *Icarus' Second Chance: The Basis and Perspectives of Space Ethics*, New York: Springer.

Aron, J. (2016) "Inflatable Space Hotel to be Tested by Space Station Crew," *New Scientist*, April 4. https://www.newscientist.com/article/2083006-inflatable-space-hotel-to-be-tested-by-space-station-crew/ (accessed August 1, 2016).

Arthur, K. G. (2012), "Descent, Elevation and Ascent: Oppositional Forces in the Strozzi di Mantova Chapel," in M. D. Edwards and E. Beiley (eds.), *Gravity in Art: Essays on Weight and Weightlessness in Painting, Sculpture and Photography*, London: McFarland, pp. 50–71.

Arthur, R. (2016), "The Future of Fashion: 10 Wearable Tech Brands You Need to Know," *Forbes*, June 30. https://www.forbes.com/sites/rachelarthur/2016/06/30/the-future-of-fashion-10-wearable-tech-brands-you-need-to-know/#422083864220 (accessed March 10, 2017).

Ashford, D. (2002), *Spaceflight Revolution*, London: Imperial College Press.

Aspelund, K. (2017), "Textile Needs and Constraints for Intravehicular Activity in Long Duration Space Flight. Part 1: Weight Limitations and Hygiene." *AATCC Review*, vol. 17, no. 2, pp. 45–52.

Badminton, N. (2004), *Alien Chic: Posthumanism and the Other Within*, New York: Routledge.

Baldaia, S. (2015), "Space Age Fashion," *Berg Fashion Library*, London: Berg.

Ballantine (2015a), "How Do You Make a Whisky Glass for Space?" *Medium*, September 2. https://medium.com/space-glass/how-do-you-make-a-whisky-glass-for-space-33d83797ad82#.5436jx30x (accessed July 6, 2016).

Ballantine (2015b), "Introducing Ballantine's Space Glass," *Medium*, September 2. https://medium.com/space-glass/introducing-the-ballantine-s-space-glass-d7722772f9a7#.uzic01k37 (accessed July 6, 2016).

Barrett, O. (1999), "An Evaluation of the Potential Demand for Space Tourism within the United Kingdom." http://www.spacefuture.com/archive/an_evaluation_of_the_potential_demand_for_space_tourism_within_the_united_kingdom.shtml (accessed July 8, 2016).

Barschke, M. F., S. Ozkan, and M. J. Johnson (2012), "Mission Control for Citizen Space Exploration," paper presented at *SpaceOps 2012 Conference*, Stockholm, Sweden, June 11–15. http://www.spaceops2012.org/proceedings/documents/id1275539-Paper-003.pdf (accessed August 8, 2016).

BBC (2017), *In Their Element*, Series 1, Episode 2: "Oxygen: The Breath of Life," BBC Radio 4, broadcast 11 a.m. Tuesday, May 2, 2017.

Behera, B. K., and P. K. Hari (2010), *Woven Textile Structure: Theory and Applications*, Cambridge, UK: Woodhead.

Behrendsen, R. (2013), "Ergonomics and Human Factors for the Restraints and Mobility Aids on the International Space Station." https://ntrs.nasa.gov/archive/nasa/casi.ntrs.nasa.gov/20130013008.pdf (accessed February 1, 2017).

Belleau, B. D. (1987), "Cyclical Fashion Movement: Women's Day Dresses: 1860–1980," *Clothing and Textiles Research Journal*, vol. 5, no. 2, pp. 15–20.

Benaroya, H. (2016), *Turning Dust to Gold: Building a Future on the Moon and Mars*, Chichester, UK: Praxis.

Berger, E. (2016), "Virgin Galactic Returns to Flight with Its VSS Unity Spacecraft," *Arts Technica*, December 5. http://arstechnica.com/science/2016/12/virgin-galactic-returns-to-flight-with-its-vss-unity-spacecraft/ (accessed December 31, 2016).

Billings, L. (2006), "Exploration for the Masses? Or Joyrides for the Ultra-rich? Prospects for Space Tourism," *Space Policy*, vol. 22, no. 3, pp. 162–164.

Bolden, C. (2015), Letter from NASA administrator Charles Bolden to Congress, Wednesday, August 5. https://www.nasa.gov/sites/default/files/atoms/files/soyuz_seat_modification_letter.pdf (accessed June 25, 2017).

Boyle, A. (2017), "Boeing and SpaceX Win Four More NASA Orders for Space Taxis, with One Big 'If,'" *Geek Wire*, January 3. https://www.geekwire.com/2017/boeing-spacex-four-nasa-orders-space-taxis/ (accessed July 28, 2017).

Boyle, E. (2016), "A Hollywood Costume Designer Is Designing Real Spacesuits for SpaceX," Gadgette.com, May 10. http://www.gadgette.com/2016/05/10/a-hollywood-costume-designer-is-designing-real-spacesuits-for-spacex/ (accessed July 5, 2017).

Brennan, L., and A. Vecchi (2011), *The Business of Space*, New York: Palgrave Macmillan.

Breward, C. (2003), *Fashion*, Oxford: Oxford University Press.

Breward, C. (2016), *The Suit*, London: Reaktion.

Broyan, J. L., Jr., T. Schlesinger, and M. K. Ewert (2016), "Exploration Mission Benefits from Logistics Reduction Technologies," paper presented at *46th International Conference on Environmental Systems*, July 10–14, Vienna, Austria. http://ntrs.nasa.gov/archive/nasa/casi.ntrs.nasa.gov/20160003493.pdf (accessed August 15, 2016).

Bureaud, A. (2003), "The Reasons for a Symposium," *Symposium: Visibility—Legibility of Space Art*. https://www.olats.org/space/colloques/artgravitezero/t_Bureaud_en.html (accessed July 16, 2017).
Bureaud, A. (2006), " 'What if . . .' Art and Weightlessness," in M. Bakke (ed.), *Going Aerial: Air, Art, Architecture*, Maastricht, Netherlands: Jan van Eyck Academie.
Burleson, D. (2005), *Space Programs Outside the United States*, Jefferson, NC: McFarland.
Cadigan, Erin (2014), *Sourcing and Selecting Textiles for Fashion*, London: Bloomsbury.
Cadogan, D. P. (2015), "The Past and Future Space Suit," *American Scientist*, vol. 103, no. 5. http://raxional.com/the-past-and-future-space-suit/ (accessed May 14, 2017).
Carr, C., and D. Newman (2005) "When Is Running More Efficient than Walking in a Space Suit?" *SAE Technical Paper 2005-01-2970.* http://papers.sae.org/2005-01-2970/ (accessed July 25, 2017).
Carson, B. (2015), "Silicon Valley Fashion Week Is Real, and It Stars Drones and Robots instead of Models," *Business Insider*, May 11. (accessed July 25, 2016).
Carter, J. (2017), "Why the 'Ultimate Wearables' Lie in the Future of Space Exploration," *Wearable*, February 28. https://www.wareable.com/wearable-tech/space-exploration-ultimate-wearables-3985 (accessed March 10, 2017).
CBS (2011), Interview with Cady Coleman, *The Talk*, season 1, episode 6, first broadcast January 19, 2011. https://www.youtube.com/watch?v=fER9InpA7vo (accessed July 22, 2016).
Chalayan, H. (2015), Q&A with Hussein Chalayan, Alastair Spalding, and Damien Jalet, Saddler's Wells, London, October 29.
Chan, N. (2013), "How Astronauts Train Underwater," *Tested*, YouTube. https://youtu.be/BRPb0J8lZcY (accessed July 18, 2015).
Chang, Y. W., and J. S. Chern (2016), "Ups and Downs of Space Tourism Development in 60 Years from Moon Register to SpaceShip Two Crash," *Acta Astronautica* vol. 127, pp. 533–541.
Chavanne, B. H. (2008), "High Fashion," *Air & Space Magazine*, March. http://www.airspacemag.com/space/high-fashion-24618648/?no-ist (accessed August 10, 2015).
Chowdhury, A. A. (2016), "Foot Restraint Version Used during the Skylab Missions," NASA Hardware archives. https://lsda.jsc.nasa.gov/scripts/hardware/hardconfig.aspx?hardware_index=2057&hardware_id=1044 (accessed February 1, 2017).
Chua, J. M. (2017), "Buzz Aldrin Launches 'Mission to Mars' Fashion Line with Sprayground." Space.com, https://www.space.com/38988-buzz-aldrin-mission-to-mars-sprayground-fashion.html (accessed August 7, 2018).
Civil Aviation Authority (2014), *UK Government Review of Commercial Spaceplane Certification and Operations*, Technical Report. https://publicapps.caa.co.uk/docs/33/CAP1189_UK_Government_Review_of_commercial_spaceplane_certification_and_operations_technical_report.pdf (accessed August 7, 2018).
Clark, S. (2016), "NASA Spells out Support for SpaceX's Red Dragon Mars Mission," *Spaceflight Now*, July 29. https://spaceflightnow.com/2016/07/29/nasa-spells-out-support-for-spacexs-red-dragon-mars-mission/ (accessed August 3, 2016).
Clark T. K., M. C. Newman, C. M. Oman, D. M. Merfeld, and L. R. Young (2015), "Modeling Human Perception of Orientation in Altered Gravity," *Frontiers in Systems Neuroscience*, vol. 9, no. 68. http://journal.frontiersin.org/article/10.3389/fnsys.2015.00068/full (accessed June 23, 2017).
Clarke, A. C. (1968), *2001: A Space Odyssey*, London: Arrow.

Collett-White, M. (2009), "Banker-Turned-Artist Displays Zero Gravity Works," *Reuters*, January 13. https://www.reuters.com/article/us-banker-space/banker-turned-artist-displays-zero-gravity-works-idUSTRE50C42L20090113?sp=true (accessed August 7, 2018).

Collins, P., S. Kuwahara, T. Nishimura, and T. Fukuoka (1997), "Design and Construction of Zero-Gravity Gymnasium," *Journal of Aerospace Engineering*, vol. 10, no. 2, pp. 94–98.

Collins, P., R. Stockmans, and M. Maita (1995), "Demand for Space Tourism in America and Japan, and Its Implications for Future Space Activities," *Advances in the Astronautical Sciences*, vol. 91, pp. 601–610.

Collins, S. (2002), "Tate in Space," Tate Gallery. http://www2.tate.org.uk/space/default.htm (accessed August 7, 2016).

Compaing, C., and L. Devere (1855), *The Tailor's Guide: A Complete System of Cutting Every Kind of Garment to Measure, Containing Upwards of Five Hundred Diagrams*, London: Simpkin, Marshall. http://hearth.library.cornell.edu/h/hearth/browse/title/4399907.html (accessed August 25, 2016).

Concus, P., and R. Finn (1969), "On the Behavior of a Capillary Surface in a Wedge," *Applied Mathematical Science*, vol. 63, no. 2, pp. 292–299.

Connikie, Y. (2007), *The 1960s*, 2nd edn, New York: Infobase.

Crosbie, J. (2017), "Here's How Much a Ticket on a SpaceX Flight to the Moon Will Cost," Inverse.com, February 27. https://www.inverse.com/article/28424-spacex-moon-mission-ticket-cost-elon-musk (accessed November 9, 2017).

Crouch, G. I. (2001), "The Market for Space Tourism: Early Indicators," *Journal of Travel Research*, vol. 40, no. 2, pp. 213–219.

Crouch, G. I., T. M. Devinney, J. L. Louviere, and T. Islam (2009), "Modelling Consumer Behavior in Space Tourism," *Tourism Management*, vol. 30, no. 3, pp. 441–454.

CSF (2015), "Commercial Spaceflight Federation Annual Report 2014." http://www.commercialspaceflight.org/wp-content/uploads/2015/06/Annual-Report-2014.pdf (accessed January 7, 2016).

Curran, L. (1999), "An Analysis of Cycles in Skirt Lengths and Widths in the UK and Germany, 1954–1990," *Clothing and Textiles Research Journal*, vol. 17, no. 2, pp. 65–72.

Cuthbertson, A. (2016), "SpaceX Hires 'Iron Man' Costume Designer for Its Spacesuits," *Newsweek*, June 5. http://europe.newsweek.com/spacex-hires-iron-man-costume-designer-spacesuits-457293?rm=eu (accessed July 10, 2016).

Cusick, G. E. (1965), "The Dependence of Fabric Drape on Bending and Shear Stiffness," *Journal of the Textile Institute Transactions*, vol. 56, issue 11, pp. 596–606.

Dabramo, L. A. (2012), "André Courrèges," *Vogue Encyclopaedia of Designers*. http://www.vogue.it/en/news/encyclo/designers/c/andre-courreges (accessed August 26, 2016).

D'Aloia, A. (2012), 'The Intangible Ground: A Neurophenomenology of the Film Experience," *European Journal of Media Studies*, vol. 1, no. 2, pp. 219–239.

Davenport, C. (2017), "Jeff Bezos' Blue Origin Gets Closer to Flying Tourists to Space," *Los Angeles Times*, December 13. http://www.latimes.com/business/la-fi-blue-origin-launch-20171213-story.html (accessed December 15, 2017).

David, L. (2006), "Wardrobe for the Weightless: Space Inspires Fashion Show," Space.com. http://www.space.com/2479-wardrobe-weightless-space-inspires-fashion-show.html (accessed April 7, 2016).

De Monchaux, N. (2011), *Spacesuit: Fashioning Apollo*, Cambridge, MA: MIT Press.

De Visscher, E. (1993) "'There Is No Such Thing as Silence . . .': John Cage's Poetics of Silence (1991)," in Richard Kostelanetz (ed.), *Writings about John Cage*, Ann Arbor: University of Michigan Press.
Deese, R. S. (2009), "The Artifact of Nature: 'Spaceship Earth' and the Dawn of Global Environmentalism," *Endeavour*, vol. 33, no. 2, pp. 70–75.
Dickens, P. (2009), "The Cosmos as Capitalism's Outside," in D. Bell and M. Parker (eds.), *Space Travel & Culture: From Apollo to Space Tourism*, Malden, MA: Blackwell, pp. 66–82.
Diliwar, A. (2015), "Introducing the Astronaut Clothing of the Future," *Newsweek*, July 5. http://www.newsweek.com/2015/07/17/introducing-astronaut-clothing-future-349955.html (accessed August 10, 2015).
Dinerman, T. (2015), "2016 Could Be the Year Space Tourism Takes Off," *The Observer*, September 22. http://observer.com/2015/09/2016-could-be-the-year-space-tourism-takes-off/ (accessed January 9, 2016).Dominoni, A. (2003), "Aesthetics in Microgravity," in J. E. Katz (ed.), *Machines That Become Us: The Social Context of Personal Communication Technology*, New Brunswick, NJ: Transaction, pp. 277–284.
Dominoni, A. (2005), "VEST. Clothing Support System On-Orbit Validation," paper presented at the 25th International Conference on Environmental Systems (ICES), Rome, Italy, July 11–14. http://hdl.handle.net/11311/255047 (accessed July 25, 2015).
Dominoni, A. (2015), *For Designers with Their Head beyond the Clouds* [ebook], Santarcangelo di Romagna: Maggioli.
Dominoni, A. (2016), "Astronaut Postures—Fashion Gestures," *Couture in Orbit*, European Space Agency. http://www.esa.int/Our_Activities/Human_Spaceflight/Couture_in_orbit/Astronaut_postures_fashion_gestures (accessed August 9, 2016).
Doule, O. (2014a), "Ground Control: Space Architecture as Defined by Variable Gravity," *Architectural Design*, vol. 84, no. 6, pp. 90–95.
Doule, O. (2014b), "Space Tourism: Waiting for Ignition," *Architectural Design*, vol. 84, no. 6, pp. 64–69.
Druyan, A. (2013), "Reflections on the Pale Blue Dot," *NASA Jet Propulsion Laboratory*, YouTube, February 13. https://www.youtube.com/watch?v=d7aL0ZGjoeg (accessed August 8, 2016)
Dubois, K. (1994), "Dance and Weightlessness: Dancers' Training and Adaptation Problems in Microgravity," *Leonardo*, vol. 27, no. 1, pp. 57–64.
Dubois, K. (2001), "Dance and Weightlessness" ("*Danse et apesanteur*"), in E. Chardronnet (ed.), *Exit Gravity* ("Quitter la gravité"), Association des Astronautes Autonomes. http://www.lyber-eclat.net/lyber/aaa/danse.html (accessed August 4, 2016).
Duda, K. (2014), "Variable Vector Countermeasure Suit (V2Suit) for Space Habitation and Exploration," Phase II Final Report, *NASA Innovative Advanced Concepts (NIAC)*. https://www.nasa.gov/sites/default/files/files/Duda_2012_PhII_V2Suit.pdf (accessed March 20, 2017).
Dunbar, B., and J. I. Petty (eds.) (2003), "Astronauts' Dirty Laundry," *NASAexplores*. http://www.nasa.gov/vision/space/livinginspace/Astronaut_Laundry.html (accessed August 15, 2016).
Dunn, M. (2016), "A Year in Space: Aching Joints and the Weird Feeling of Clothing Touching Your Skin," *Global News*, March 4. https://globalnews.ca/news/2558552/a-year-in-space-aching-joints-and-the-weird-feeling-of-clothing-touching-your-skin/ (accessed January 27, 2018).

Edwards, M. D. (2012), "Altichiero, Giotto, Dante, and the Metaphorical Use of Gravity and Levity," in M. D. Edwards and E. Bailey (eds.), *Gravity in Art: Essays on Weight and Weightlessness in Painting, Sculpture and Photography*, London: McFarland, pp. 72–85.

Edwards, M. D., and E. Bailey (2012), "Introduction," in M. D. Edwards and E. Bailey (eds.), *Gravity in Art: Essays on Weight and Weightlessness in Painting, Sculpture and Photography*, London: McFarland, pp. 1–9.

Enders, C. (2015), "Space Travel for the 1%: Virgin Galactic's $250,000 Tickets Haunt New Mexico Town," *The Guardian*, October 4. http://www.theguardian.com/us-news/2015/oct/04/space-travel-virgingalactic-haunts-new-mexico-town (accessed January 7, 2016).

ESA (2016), "ExoMars Sets Sights on the Red Planet," *European Space Agency*, June 16. http://exploration.esa.int/mars/57943-exomars-sets-sights-on-the-red-planet/ (accessed August 3, 2016).

Eshun, K. (2005), "On the Use and Abuse of Microgravity for Life," in A. C. T. Geppert (ed.), *Imagining Outer Space: European Astroculture in the Twentieth Century*, New York: Palgrave MacMillan, pp. 28–31.

Evans, C., and S. Frankel (2008), *The House of Viktor & Rolf*, London: Merrell.

Evans, C. (2007), *Fashion at the Edge: Spectacle, Modernity and Deathliness*, 2nd edn, London: Yale University Press.

Fassl, J. (2012), "The Suspension of Gravity in Giandomenico Tiepolo's *Punchinelli on a Swing*," in M. D. Edwards and E. Bailey (eds.), *Gravity in Art: Essays on Weight and Weightlessness in Painting, Sculpture and Photography*, London: McFarland, pp. 137–146.

Fecht, S. (2016), "As Boeing Slips, SpaceX Will Likely Be First Company to Carry Astronauts," *Popular Science*, May 13. http://www.popsci.com/spacex-will-likely-be-first-private-company-to-carry-astronauts-into-space (accessed August 1, 2016).

Ferl, L., L. Hewes, D. Cadogan, D. Graziosi, and K. Splawn (2006), "System Considerations for an Exploration Spacesuit Upper Torso Architecture," paper presented at 36th International Conference on Environmental Systems, Norfolk, Virginia, July 17–20. http://papers.sae.org/2006-01-2141/ (accessed August 15, 2015).

Fernholz, T. (2015), "The Space Suit You Could Be Wearing a Few Years from Now Is Being Designed in a Small Workshop in Brooklyn," *Quartz*, March 17. http://qz.com/356649/the-space-suit-you-could-be-wearing-a-few-years-from-now-is-being-designed-in-a-small-workshop-in-brooklyn/ (accessed October 12, 2015).

Ferreira, B. (2015), "Space Station Chic: How NASA Is Engineering Better Indoor Clothes for Astronauts," *Motherboard*, June 24. http://motherboard.vice.com/read/space-station-chic-how-nasa-is-engineering-better-indoor-clothes-for-astronauts (accessed August 10, 2015).

FFD (2012), "Final Frontier Design's 3G Space Suit," Kickstarter campaign. https://www.kickstarter.com/projects/872281861/final-frontier-designs-3g-space-suit/description (accessed May 14, 2017).

FFD (2017), "Space Suit Experience," Final Frontier Design. http://www.finalfrontierdesign.com/space-suit-experience/ (accessed May 29, 2017).

Fiell, C., and P. Fiell (2005), *Design of the 20th Century*, Köln: Taschen.

Figueiredo Neto, A. M., and S. R. A. Salinas (2005), *The Physics of Lyotropic Liquid Crystals: Phase Transitions and Structural Properties*, Oxford: Oxford University Press.

Finnegan, E. (2014), "Zero Gravity Wedding," *Vimeo*. https://vimeo.com/103361634 (accessed August 9, 2015).

Fisher, A. (2009), " 'Very Stunning and Very Cool': The Launch of Space Tourism," *Technology Review*, Jan/Feb, pp. 58–69.

Fleischer, T. L. (2015), "Falling for Lavin," *La Mode* [blog]. http://tanjafleischer.weebly.com/fashion/falling-for-lanvin (accessed July 16, 2016).

Forrester, I. (1908), "The Marvelous Hats of 1908," *Munsey's Magazine*, vol. 39, pp. 519–524.

Fratto, L. (2005), "Evaluating the Performance of Fibres and Fabrics for Astronaut's IVA Wardrobes for Long Duration Space Flight," MS thesis, North Carolina State University. http://repository.lib.ncsu.edu/ir/bitstream/1840.16/2074/1/etd.pdf (accessed August 10, 2015).

Freeland, S. (2005), "Up, Up and Back: The Emergence of Space Tourism and Its Impact on the International Law of Outer Space," *Chicago Journal of International Law*, vol. 6, no. 1, pp. 1–22.

Friedman, L. (2013), "The New Space Race: It's Not Just the USA Anymore," *Los Angeles Times*, December 9. http://articles.latimes.com/2013/dec/09/opinion/la-oe-friedman-new-space-race-20131209 (accessed January 7, 2016).

Frost, R. (2012), "What Does It Sound Like Inside a Space Suit, while in Space?" *Quora*, November 23. http://www.quora.com/What-does-it-sound-like-inside-a-space-suit-while-in-space (accessed June 20, 2015).

Frost, R. (2015), "Why It's Impossible to Design a Perfect Spacesuit," *Gizmodo*, February 2. http://gizmodo.com/why-its-impossible-to-design-a-perfect-spacesuit-1683283611 (accessed June 9, 2015).

FUTRON Corporation (2006), "Suborbital Space Tourism Demand Revisited," Futron, Bethesda.

Ganz München (2002), "Esprit Megastore Opening München: Vertical Catwalk begeisterte in der Fußgängerzone," *Ganz München*. http://www.ganz-muenchen.de/shopping/mode/esprit/vertical_catwalk/info.html (accessed July 17, 2016).

Garriott, R. (2008), "Art in Space," *Our Space*. http://www.our-space.org/materials/states-of-matter/art-in-space (accessed January 23, 2017).

Gast, M. A. (2010), "Commercial Spacewalking: Designing an EVA Qualification Program for Space Tourism," *NASA Technical Report Server.* http://ntrs.nasa.gov/archive/nasa/casi.ntrs.nasa.gov/20100011328.pdf (accessed August 15, 2015).

Gast, M. A., and S. K. Moore (2011), "A Glimpse from the Inside of a Space Suit: What Is It Really like to Train for an EVA?" *Acta Astronautica*, vol. 68, no. 1–2, pp. 316–325.

Geppert, A. C. T. (2012), *Imagining Outer Space: European Astroculture in the Twentieth Century*, New York: Palgrave MacMillan.

Geppert, A. C. T. (ed.) (2018), *Limiting Outer Space Astroculture After Apollo*, New York: Palgrave MacMillan.

Geran, M. (2006), "It's a Man's World," *Interior Design Magazine* [online], 4.01. http://kramerdesigngroup.com/press/intdes4.04_presskit.pdf (accessed July 12, 2010).

Gibson, D. (2012), *Commercial Space Tourism: Impediments to Industrial Development and Strategic Communication Solutions*, Oak Park, IL: Bentham.

Gittleson, K. (2013), "The Design Firms Pioneering Space Suit Fashion," *BBC News* [online], September 23. http://www.bbc.co.uk/news/business-23947790 (accessed May 28, 2017).

Gleason, K. (2012), *Alexander McQueen: Evolution*, New York: Race Point.

Gooden, B. (2012), *Spacesuit: A History through Fact and Fiction*, Pulborough: Tattered Flag.

Gorman, T. (2015), "This Is How You Get Wasted in Space," Interview with Samuel Coniglio, *Ozy*, October 28. http://www.ozy.com/good-sht/this-is-how-you-get-wasted-in-space/33113 (accessed July 6, 2016).

Graziosi, D., J. Ferl, and K. Splawn (2004), "An Examination of Spacesuit Entry Types and the Effect on Suit Architecture," *Proceedings of the AIAA Space Conference and Exposition,* San Diego, California, pp. 1229–1237.

Graziosi, D., J. Ferl, and K. Splawn (2005), "Evaluation of a Rear Entry System for an Advanced Spacesuit," SAE Technical Paper 2005-01-2976. http://papers.sae.org/2005-01-2976/ (accessed July 25, 2017).

Groom, A. (2011), "Luxury in Fashion Reconsidered," *Fashion Theory*, vol. 15, no. 4, pp. 503–508.

Grzinic, M. (2003), "Aesthetics of the Digitized Body," in Kiyokazu Nishimura (ed.), *Selected Papers of the 15th International Congress of Aesthetics*, pp. 81–86.

Guillaume, V. (1998), *Fashion Memoir: Courrèges*, London: Thames and Hudson.

Hadfield, C. (2014), "What I Learned from Going Blind in Space," *TED*. https://www.ted.com/talks/chris_hadfield_what_i_learned_from_going_blind_in_space/transcript?language=en (accessed July 18, 2015).

Harrington, R. (2016), "A Famous Astronaut Says His Body Still Hurts after Spending a Year in Space," *Tech News*, June 8. http://uk.businessinsider.com/scott-kelly-space-travel-effects-human-body-2016-6?r=US&IR=T (accessed December 15, 2017).

Hawking, S. (2016), "I Think the Human Race Has No Future If It Doesn't Go to Space," *The Guardian*, September 26, extract from "Afterword: Space Here I Come," in J. Guthrie (2016), *How to Make a Spaceship*, London: Bantam Press, pp. 413–416. https://www.theguardian.com/science/2016/sep/26/i-think-the-human-race-has-no-future-if-it-doesnt-go-to-space (accessed May 27, 2017).

Hayles, N. K. (1999), *How We Became Posthuman*, Chicago: University of Chicago Press.

Helmreich, S. (2011), "From Spaceship Earth to Google Ocean: Planetary Icons, Indexes, and Infrastructures," *Social Research*, vol. 78, no. 4, pp. 1211–1242.

Herbert, D. (dir.) (1999), "Space Wear," *That NASA Show*, episode 2, Show B, Version D, https://archive.org/details/space_wear (accessed August 9, 2015).

Hersch, M. H. (2009a), "Checklist: The Secret Life of Apollo's 'Fourth Crewmember,' " in D. Bell and M. Parker (eds.), *Space Travel & Culture: from Apollo to Space Tourism*, Malden, MA: Blackwell, pp. 6–24.

Hersch, M. H. (2009b) "High Fashion: The Women's Undergarment Industry and the Foundations of American Spaceflight," *Fashion Theory*, vol. 13, no. 3, pp. 345–370.

Hitt, D., O. K. Garriott, and J. Kerwin (2008), *Homesteading Space: The Skylab Story*, Lincoln: University of Nebraska Press.

Holick, M. F. (2000), "Microgravity-Induced Bone Loss—Will It Limit Human Space Exploration?" *The Lancet*, vol. 355, no. 9215, pp. 1569–1570. http://www.thelancet.com/pdfs/journals/lancet/PIIS0140-6736(80)02208-8.pdf (accessed July 19, 2017).

Holpuch, A. (2014), "Nasa Says New Spacesuit One Small Step towards Sending Mankind to Mars," *The Guardian* [online]. http://www.theguardian.com/science/2014/apr/30/nasa-spacesuit-zseries-new-design-mars (accessed April 30, 2014).

Horwell, V. (2016), "André Courrèges Obituary," *The Guardian* [online], January 8. https://www.theguardian.com/fashion/2016/jan/08/andre-courreges (accessed August 26, 2016).

Howarth, D. (2016), "Y-3 Creates Spacesuits for Virgin Galactic Pilots on World's First Commercial Space Flights," *Dezeen*, January 14. https://www.dezeen.com/2016/01/14/y-3-adidas-spacesuits-virgin-galactic-worlds-first-commercial-space-flights-video-movie/ (accessed January 1, 2017).

Howell, E. (2013), "Earth Living Is Tough for Astronaut Used to Space," Space.com. http://www.space.com/21413-hadfield-astronaut-health-return-earth.html (accessed February 12, 2017).

Howell, E. (2015), "Mock Mars Mission Crew Tests 'Smart Shirt' for Astronauts," Space.com, March 27. http://www.space.com/28945-mock-mars-mission-hexoskin-shirt.html (accessed March 20, 2017).

Howell, E. (2016), "SpaceShipTwo: On a Flight Path to Space Tourism," Space.com, February 17. http://www.space.com/19021-spaceshiptwo.html (accessed December 31, 2016).

Howell, E. (2017a), "SpaceX: First Private Flights to Space Station," Space.com, February 19. http://www.space.com/18853-spacex.html (accessed January 7, 2016).

Howell, E. (2017b), "Augmented Footwear for Prehensile Capability in Microgravity Environments," White Paper.

Howell, E. L. (2011), "Augmented Footwear for Gripping and Holding in Micro-gravity Environments," US Patent 20110296715 A1. http://dx.doi.org/10.2514/6.2011-5032 (accessed May 14, 2017).

Hu, J. (2004), *Structure and Mechanics of Woven Fabrics*, Cambridge, UK: Woodhead.

Hunter, L., and J. Fan (2008), "Measuring and Predicting Fabric and Garment Drape," in Catherine Fairhurst (ed.), *Advances in Apparel Production*, Cambridge, UK: Woodhead, pp. 7–25.

Iannotta, B. (1997), "Learning from Mir's Mishaps," *Aerospace America*, vol. 35, no. 11, pp. 32–36.

Igenbergs, E., W. Naumann, and E. Pfeiffer (1996), "Munich Space Chair (MSC): A New Body Restraint System for Astronauts," Proceedings of the *Space Station Utilisation* Symposium, Darmstadt, Germany, September 30–October 2, pp. 129–131. http://adsabs.harvard.edu/full/1996ESASP.385..129I (accessed February 12, 2017).

Impey, C. (2015), "There Will Be Sex in Space: Mars, Science and 'the Dictates of Biology and Human Culture,'" *Salon*, April 12. http://www.salon.com/2015/04/12/there_will_be_sex_in_space_mars_science_and_the_dictates_of_biology_and_human_culture/ (accessed February 12, 2017).

Incredible Adventures (2016), "Lose Weight the Fun Way—No Astronaut Experience Required." http://www.incredible-adventures.com/zero-gravity-usa.html (accessed December 31, 2016).

Ivanovich, G. S. (2008), *Salyut the First Space Station: Triumph and Tragedy*, Chichester, UK: Springer/Praxis.

JAXA (2014), "Koichi Wakata Weekly Activity Report," Vol. 19, YouTube. https://www.youtube.com/watch?v=MlqXsl7kauo (accessed July 29, 2017).

Jenkin Jones, S. (2011), *Fashion Design*, London: Laurence King.

Jiang, Y., M. Cui, and J. Hu (2012), "Computer Generation of 3D Textile Draping Simulation," proceedings of the AASRI Conference on Modeling, Identification and Control, pp. 661–666.

Johnston, N. (2012), "Beneath Sci-fi Sound: *Primer*, Science Fiction Sound Design, and American Independent Cinema," *Alphaville: Journal of Film and Screen Media*, vol. 3. https://cora.ucc.ie/bitstream/handle/10468/1450/ArticleJohnston.pdf (accessed June 24, 2015).

Jordan, J. W. (2003), "Kennedy's Romantic Moon and Its Rhetorical Legacy for Space Exploration," *Rhetoric & Public Affairs*, vol. 6, no. 2, pp. 209–231.

Jordan, N. C., J. H. Saleh, and D. J. Newman (2006), "The Extravehicular Mobility Unity: A Review of Environment, Requirement, and Design Changes in the US Spacesuit," *Acta Astronautica*, vol. 59, no. 12, pp. 1135–1145.

Jorgensen, D. (2009), "Middle America, the Moon, the Sublime and the Uncanny," in D. Bell and M. Parker (eds.), *Space Travel & Culture: From Apollo to Space Tourism*, Malden, MA: Blackwell, pp. 178–203.

Kac, E. (2005), "Against Gravitropism: Art and the Joys of Levitation," in N. Triscott and R. La Frenais (eds.), *Zero Gravity: A Cultural User's Guide*, London: Arts Catalyst, pp. 18–25.

Kaminski, A. (2016), "What Place for People: The Role of the Public and NGOs in Space Innovation and Governance," in C. Al-Ekabi, B. Baranes, P. Hulsroj, and A. Lahcen (eds.), *Yearbook on Space Policy 2014: The Governance of Space*, New York: Springer, pp. 217–228.

Karmali, F., and M. Shelhamer (2008), "The Dynamics of Parabolic Flight: Flight Characteristics and Passenger Percepts," *Acta Astronautica*, vol. 63, no. 5–6, pp. 594–602. http://doi.org/10.1016/j.actaastro.2008.04.009 (accessed December 31, 2016).

Katayama, L. (2007), "The Cosmos Meets Cosmo," *Wired*, April 1. https://www.wired.com/2007/04/the-cosmos-meets-cosmo/ (accessed June 25, 2017).

Kelly, S. (2016), Q&A with Astronaut Scott Kelly, *Reddit*, January 23. https://www.reddit.com/r/IAmA/comments/42cllg/i_am_astronaut_scott_kelly_currently_spending_a/cz9aepd/ (accessed February 12, 2017).

Kempner, M. J. (1947), "Travel Directions: Formula for Flying," *Vogue*, January 1, pp. 111, 146.

Kemsley, T. (2013) "Bacteria in Space Behave Differently than on Earth: A Study," *Nature World News*, July 10. http://www.natureworldnews.com/articles/2889/20130710/bacteria-space-behaves-differently-earth-study.htm (accessed August 18, 2016).

Kennedy, J. F. (1962), "Moon Speech," presented at Rice University stadium, September 12. http://er.jsc.nasa.gov/seh/ricetalk.htm (accessed August 9, 2016).

Kitney, D. (2017), "Virgin's Space Flights Fully Booked until 2021," *Australian Business Review*, May 19. http://www.theaustralian.com.au/business/aviation/virgins-space-flights-fully-booked-until-2021/news-story/5171757ae21cb9916f59e872a09624cf (accessed July 19, 2017).

Kleanthous, D. (2016), "Y-3 & Virgin Galactic—The Space Suit," Nomadic Films. http://www.dreykleanthous.com/y3virgingalacticthespacesuit (accessed January 1, 2017).

Klotz, I. (2012), "NASA Mandates Pressurized Spacesuits for Commercial Crew Fliers," *Space News*, November 26. http://spacenews.com/nasa-mandates-pressurized-spacesuits-for-commercial-crew-fliers/ (accessed May 27, 2017).

Kluger, J. (2009), "How Serious a Setback Is the SpaceX Rocket Explosion?" *Time* [online], June 28. http://time.com/3939217/spacex-rocket-explosion-setback/ (accessed August 1, 2016).

Kohonen, I. (2009), "The Space Race and Soviet Utopian Thinking," in D. Bell and M. Parker (eds.), *Space Travel & Culture: From Apollo to Space Tourism*, Malden, MA: Blackwell, pp. 114–131.

Kolodziej, M. (2012), "From Vertigo to Ethereality in Environmental Art," in M. D. Edwards and E. Bailey (eds.), *Gravity in Art: Essays on Weight and Weightlessness in Painting, Sculpture and Photography*, London: McFarland, pp. 293–303.

Kossoff, M (2016), "SpaceX Rocket Explodes and Destroys Facebook's First Satellite," *Vanity Fair*, September 1. http://www.vanityfair.com/news/2016/09/spacex-rocket-explodes-facebook-satellite (accessed February 1, 2017).

Kramer, M. (2014a), "Suit Up! Final Frontier Design Launches Space Suit Experience in NYC," Space.com, August 29. http://www.space.com/26977-final-frontier-design-spacesuit-experience.html (accessed May 29, 2017).

Kramer, M. (2014b), "What to Wear in Space: Spacesuit Chic with Final Frontier Design," Space.com, August 28. http://www.space.com/26947-final-frontier-design-spacesuit-interview.html (accessed May 27, 2017).

Kramer, M. (2015), "NASA's Next Space Race: SpaceX vs. Boeing," Space.com, January 30. http://www.space.com/28411-spacex-boeing-nasa-space-race.html (accessed January 7, 2016).

Kuijpers, A., and R. Gong (2014), "Virtual Tailoring for Enhancing Product Development and Sales," paper presented at Fashion Colloquia 2014, Amsterdam, Netherlands, January 23–24. http://gfc-conference.eu/files_download/GFC2014/KUIJPERS&GONG_Virtual_tailoring_for_enhancing_product_development_and_sales.pdf (accessed November 17, 2016).

Kumar, P. (2012), *Product Design: Creativity, Concepts and Usability*, New Delhi: PHI Learning.

La Frenais, R. (2005), "An Introduction to Vertigo," in A. C. T. Geppert (ed.), *Imagining Outer Space: European Astroculture in the Twentieth Century*, New York: Palgrave MacMillan, pp. 8–10.

Landfest, U., N. Remuss, K. Schrogl, and J. Worms (2011), *Humans in Outer Space: Interdisciplinary Perspectives*, Springer: New York.

Lathers, M. (2012), *Space Oddities: Women and Outer Space in Popular Film and Culture, 1960–2000*, New York: Continuum.

Launius, R. D. (2008), "Heroes in a Vacuum: The Apollo Astronaut as Cultural Icon," *Florida Historical Quarterly*, vol. 87, no. 2, pp. 174–209.

Lazier, B. (2011), "Earthrise; or, the Globalization of the World Picture," *American Historical Review*, vol. 116, no. 3, pp. 602–630.

Lee, G., D. Graziosi, K. Splawn, J. and Ferl (2011), "Enhancing Capability in Launch, Entry, and Abort Style Spacesuits For ISS and Commercial Use," proceedings of the 41st International Conference on Environmental Systems, Portland, Oregon, July 17–21.

Lipson, H., and M. Kurman (2013), *Fabricated: The New World of 3D Printing*, Indianapolis, IN: Wiley.

Lockard, E. (2014), *Human Migration to Space: Alternative Technological Approaches for Long-Term Adaptation to Extraterrestrial Environments*, New York: Springer.

Loehr, J. A., S. M. Lee, K. L. English, J. D. Sibonga, S. M. Smith, B. A. Spiering, and R. D. Hagan (2011), "Musculoskeletal Adaptations to Training with the Advanced Resistive Exercise Device," *Medicine and Science in Sports and Exercise*, vol. 43, no. 1, pp. 146–156.

Logsdon, J. M. (2010), *John F. Kennedy and the Race to the Moon*, New York: Palgrave Macmillan.

Loschek, I. (2009), *When Clothes Become Fashion: Design and Innovation Systems*, Oxford: Berg.

Lubitz, R. (2016), “These Are the Designers Working Hard to Make Stylish Clothes for Wheelchair Users,” Style.Mic. https://mic.com/articles/132453/these-are-the-designers-working-hard-to-make-stylish-clothes-for-wheelchair-users#.ckeGOhKhy (accessed January 14, 2017).

Lundén, E. C. (2016), “Barbarella’s Wardrobe: Exploring Jacques Fonteray’s Intergalactic Runway,” *Film, Fashion & Consumption*, vol. 5, no. 2, pp. 185–211.

Mackenzie, M. (2009) . . . *Isms: Understanding Fashion*, London: Herbert Press.

Mackey, A. (2014), “Spacewear,” *Angella Mackey: Hyper-functional Clothing Design*. http://www.angellamackey.com/portfolio/?portfolio=space-travel (accessed January 24, 2017).

Mackrell, A. (2005), *Art and Fashion*, London: B. T. Batsford.

McKinnon, M. (2015), “Astronauts Have Done So, So Much with Duct Tape and Electrical Tape,” *Gizmodo*, June 17. http://gizmodo.com/astronauts-have-done-so-so-much-with-duct-tape-and-ele-1711503831 (accessed July 14, 2017).

McRae, L. (2016), “Prepping the Body for Space.” http://www.lucymcrae.net/prepping-the-body-for-space/ (accessed August 10, 2016).

Meggs, L. (2015), Interview with Evelyn Orndoff and Darwin Poritz, in *Space Station Live: What the Well-Dressed Astronaut is Wearing*, NASA Johnson. https://www.youtube.com/watch?v=NdL44q6mAuE (accessed August 21, 2016).

Mei, Z., W. Shen, Y. Wang, J. Yang, T. Zhou, and H. Zhou (2015), “Unidirectional Fabric Drape Testing Method,” *PLoS ONE*, vol. 11, no. 10, pp. 1–15.

Menkes, S. (2000), “Fashion’s Poet of Black: Yamamoto,” *New York Times* [online], September 5. http://www.nytimes.com/2000/09/05/style/fashions-poet-of-black-yamamoto.html (accessed January 1, 2017).

Messier, D. (2011), “Commercial Spacesuit Companies Compete for Market Share,” Parabolic Arc, March 21. http://www.parabolicarc.com/2011/03/21/commercial-spacesuit-companies-compete-market-share/ (accessed May 28, 2017).

Messier, D. (2014), “Final Frontier Design launches ‘Space Suit Experience,’ ” Parabolic Arc, August 29. http://www.parabolicarc.com/2014/08/29/final-frontier-design-launches-space-suit-experience/ (accessed May 29, 2017).

Moeran, B. (2004), “Women’s Fashion Magazines: People, Things, and Values,” in C. Werner and D. Bell (eds.), *Value and Valuables: From the sacred to the Symbolic*, Oxford, UK: Altamira, pp. 257–281.

Montgomery, E. (1948), “Around-the-World Wardrobe: 66 Pounds and 60 Days,” *Vogue*, May, pp. 134–136.

Monteith, S. (2008), *American Culture in the 1960s*, Edinburgh: Edinburgh University Press.

Murray, C. D., and J. Sixsmith (1999), “The Corporeal Body in Virtual Reality,” *Ethos*, vol. 27, no. 3, pp. 315–343.

Murray, J., A. Waldie, and D. J. Newman (2014), *Gravity-Loading Body Suit*. US Patent 8769712 B2. https://www.google.com/patents/US8769712 (accessed May 7, 2017).

Musk, E. (2015), Reddit Q&A, January 5, 2015. https://www.reddit.com/r/IAmA/comments/2rgsan/i_am_elon_musk_ceocto_of_a_rocket_company_ama (accessed July 24, 2015).

Musk, E. (2016), “Making Humans a Multiplanetary Species,” presentation at the 67th International Astronautical Conference, Guadalajara, Mexico, September 26–30.

Musk, E. (2017), “Making Humans a Multi-Planetary Species,” *New Space*, vol. 5, No. 2, pp. 45–61.

NASA (2002), "Waste Collection System." https://spaceflight.nasa.gov/shuttle/reference/shutref/orbiter/eclss/wcs.html (accessed February 12, 2017).

NASA (2007), "International Space Station Catalogue of Intravehicular Activity (IVA) Government Furnished Equipment (GFA) Flight Crew Equipment (FCE)." jsc-28533e international space station catalogue activity – EverySpec (accessed July 5, 2017).

NASA (2010), "NASA Open Government Plan," United States Government National Aeronautics and Space Administration.

NASA (2011), "Higher Altitude Improves Station's Fuel Economy." http://www.nasa.gov/mission_pages/station/expeditions/expedition26/iss_altitude.html (accessed August 4, 2016).

NASA (2014a), "Flight Safety Data Package—Phase I," template. https://www.nasa.gov/sites/default/files/atoms/files/template_phase_i_sdp_0.docx (accessed July 16, 2017).

NASA (2014b), "The NASA Z-2 Suit." http://jscfeatures.jsc.nasa.gov/z2/ (accessed May 1, 2014).

NASA (2014c), "Space Station 3-D Printer Builds Ratchet Wrench to Complete First Phase of Operations." https://www.nasa.gov/mission_pages/station/research/news/3Dratchet_wrench (accessed March 14, 2017).

NASA (2014d), "Z-2 Prototype Suit Cover Layer Designs." https://www.nasa.gov/content/z-2-prototype-suit-cover-layer-designs-0/#.WGhnQ6KLRR1 (accessed January 1, 2017).

NASA (2015a), "International Space Station Tour." http://www.nasa.gov/mission_pages/station/main/suni_iss_tour.html (accessed July 16, 2016).

NASA (2015b), "NASA's Journey to Mars: Pioneering Next Steps in Space Exploration." http://www.nasa.gov/sites/default/files/atoms/files/journey-to-mars-next-steps-20151008_508.pdf (accessed October 28, 2015).

NASA Johnson Space Center (2013), "NASA Astronaut Karen Nyberg Invites Quilters to Create a Space Square," October 31. https://www.youtube.com/watch?v=N0c1CijmH7o (accessed July 19, 2017).

NASA Technical Standard (2015), "NASA Space Flight Human-System Standard, Volume 2: Human Factors, Habitability, and Environmental Health (Revision A)." file:///Users/aatqbkb/Downloads/NASA-STD-3001-VOL--2A.pdf (accessed August 15, 2016).

Newman, D. (2012), "Building the Future Spacesuit," *Ask Magazine*, no. 45, pp. 37–40. https://www.nasa.gov/pdf/617047main_45s_building_future_spacesuit.pdf (accessed July 22, 2017).

Newman, D. (2013), "Smart Suits to Enable Astronaut Exploration of Mars," lecture presented at the Radcliffe Institute, Harvard University, November 21. https://www.radcliffe.harvard.edu/event/2013-dava-j-newman-lecture (accessed August 16, 2015).

Newman, D. (2014), "Building the Future Spacesuit," lecture presented at the Harvard-Smithsonian Center for Astrophysics, May 15. https://www.youtube.com/watch?v=7hkzvoOJA5E (accessed August 14, 2015).

Noble Biomaterials (2016), "Frequently Asked Questions," *X-Static Performance Textiles*. http://noblebiomaterials.com/xstatic-textiles/faq/ (accessed August 15, 2016).

Norris, C. and Bainbridge, J. (2009), "Selling Otaku? Mapping the Relationship between Industry and Fandom in the Australian Cosplay Scene," *Intersections: Gender and Sexuality in Asia and the Pacific*, no. 20. http://intersections.anu.edu.au/issue20/norris_bainbridge.htm (accessed May 29, 2017).

Nowack, E. (1999), "Ergonomics and Rehabilitation," in Waldemar Karwowski and William S. Marras (eds.), *The Occupational Ergonomics Handbook*, New York: CRC Press, pp. 1333–1352.

Nowack, E. (2001), "Anthropometry for Design," in Waldemar Karwowski (ed.), *The Encyclopedia of Ergonomics and Human Factors, Volume II*, New York: Taylor and Francis, pp. 875–878.

O'Callaghan, J. (2015), "Could This $300 Billion 'Space Mushroom' Replace the ISS? Giant Rotating Station Would Create Artificial Gravity for Astronauts," *Daily Mail* [online], April 8. http://www.dailymail.co.uk/sciencetech/article-3030087/Could-300-billion-space-mushroom-replace-ISS-Giant-rotating-station-create-artificial-gravity-astronauts.html (accessed July 14, 2016).

Oberg, J. (1998), "Shuttle-Mir's Lessons for the International Space Station," *IEEE Spectrum*, vol. 35, no. 6, pp. 28–37.

Oberhaus, D. (2015a), "Making A Sexier Spacesuit," *Slate*, October 7. http://www.slate.com/articles/technology/future_tense/2015/10/creating_sexier_spacesuits_for_the_commercial_space_race.html (accessed May 14, 2017).

Oberhaus, D. (2015b), "Meet the Hollywood Prop Designer Making NASA's Spacesuits," *Vice*, September 2. https://www.vice.com/en_uk/article/meet-the-hollywood-prop-designer-making-nasas-spacesuits-511 (accessed May 14, 2017).

Okonkwo, U. (2016), *Luxury Fashion Branding: Trends, Tactics, Techniques*, Basingstoke: Palgrave Macmillan.

Orndoff, E. (2013), "IVA Clothing Study," paper presented at the Joint Science Symposium, Johnson Space Center, Houston, December 10–12. http://ntrs.nasa.gov/archive/nasa/casi.ntrs.nasa.gov/20140004221.pdf (accessed August 21, 2016).

Orndoff, E. (2015a), "Space Wear Vision: Development of a Wardrobe for Life in Space Vehicles and Habitats," NASA Technical Reports. http://ntrs.nasa.gov/archive/nasa/casi.ntrs.nasa.gov/20150003471.pdf (accessed February 7, 2016).

Orndoff, E. (2015b), "Space Wear Vision," NASA Technical Reports. http://ntrs.nasa.gov/archive/nasa/casi.ntrs.nasa.gov/20150004597.pdf (accessed August 15, 2016).

Orndoff, E. (2016), "Intravehicular Activity Clothing Study (IVA Clothing Study)," ISS Science for Everyone, NASA, July 14, http://www.nasa.gov/mission_pages/station/research/experiments/1084.html (accessed August 21, 2016).

Orndoff, E., P. Darwin, S. Thilini, and V. Byme (2014), "Pre-Flight Advanced Clothing Study," NASA Technical Reports. https://ntrs.nasa.gov/search.jsp?R=20140012991 (accessed July 19, 2017).

Overbye, D. (2006), "On the Runway: Spacewear Meant to Dazzle, Even in Zero Gravity," *New York Times* [online], May 16. http://www.nytimes.com/2006/05/16/science/16find.html?_r=0 (accessed August 9, 2015).

Palmer, J. (2005), "Achieving Levity," in N. Triscott and R. La Frenais (eds.), *Zero Gravity: A Cultural User's Guide*, London: Arts Catalyst, pp. 26–27.

Parker, M. (2009), "Capitalists in Space," in D. Bell and M. Parker (eds.), *Space Travel & Culture: From Apollo to Space Tourism*, Malden, MA: Blackwell, pp. 83–97.

Pavitt, J. (2008), *Fear and Fashion in the Cold War*, London: V&A Books.

Pearson, J. F. (1976), "Rescue in Space," *Popular Mechanics*, October, pp. 67–69.

Peeters, W. (2010), "From Suborbital Space Tourism to Commercial Personal Spaceflight," *Acta Astronautica*, vol. 66, no. 11–12, pp. 1625–1632.

Penley, C. (1997), *NASA/Trek: Popular Science and Sex in America*, London: Verso.

Pettit, D. R., M. M. Weislogel, P. Concus, and R. Finn (2011), *Beverage Cup for Drinking Use in Spacecraft or Weightless Environments*, US Patent 8074827 B2. http://www.google.co.uk/patents/US8074827 (accessed July 5, 2016).

Phillips, T. (2013), "The Zero Gravity Coffee Cup," NASA Science, *Science News*, July 15. http://science.nasa.gov/science-news/science-at-nasa/2013/15jul_coffeecup/ (accessed July 3, 2016).

Pietronigro, F. (2000), "Research Project Number 33: Investigating the Creative Process in a Microgravity Environment," *Leonardo*, vol. 33, issue 3, pp. 169–177.

Pitt, B., C. Brensinger, J. Saleh, C. Carr, P. Schmidt, and D. Newman (2001), "Astronaut Bio-Suit for Exploration Class Missions," *NIAC Phase 1 Report*. http://www.4frontiers.us/dev/assets/BioSuit-NIACPhaseIReport.pdf (accessed August 19, 2015).

Pocock, P. (2012), "Look Up! Art in the Age of Orbitization," in A. C. T. Geppert (ed.), *Imagining Outer Space: European Astroculture in the Twentieth Century*, New York: Palgrave MacMillan, pp. 319–341.

Portree, David S. F., and Robert C. Trevino (1997), "Walking to Olympus: An EVA Chronology," *Monographs in Aerospace History*, Series 7, http://spaceflight.nasa.gov/spacenews/factsheets/pdfs/EVACron.pdf (accessed August 10, 2015).

Prater, T. J., Q. A. Bean, R. D. Beshears, T. D. Rolin, N. J. Werkheiser, E. A. Ordonez, R. M. Ryan, and F. E. Ledbetter III (2016), "Summary Report on Phase I Results From the 3D Printing in Zero-G Technology Demonstration Mission, Volume I," NASA Technical Reports. https://ntrs.nasa.gov/archive/nasa/casi.ntrs.nasa.gov/20160008972.pdf (accessed January 1, 2017).

Quinn, B. (2002), *Techno Fashion*, Oxford: Berg.

Quinn, B. (2012), *Fashion Futures*, New York: Merrell.Rapp, D. (2014), "Green Is the Message," paper presented at Deception: 1st Global Conference, Mansfield College, Oxford, July 17–19. http://www.inter-disciplinary.net/probing-the-boundaries/wp-content/uploads/2014/05/DAVIDE-RAPP_wpaper-decep1.pdf (accessed August 8, 2016).

Roberts, J. (2013), *Free Cutting* [e-book]. http://www.mediafire.com/view/eabgygf9d0b9ira/FREE-CUTTING-Julian-Roberts.pdf (accessed July 14, 2016).

Roche, D., and J. Birrel (1994), *The Culture of Clothing*, Cambridge, UK: Cambridge University Press.

Ruffles, T. (2004), *Ghost Images: Cinema of the Afterlife*, London: McFarland.

Russon, M. (2016), "China Wants to Visit Mars by 2020 and Beat Nasa to Set up the First Manned Moon Base," *International Business Times* [online], April 22. http://www.ibtimes.co.uk/china-wants-visit-mars-by-2020-beat-nasa-set-first-manned-moon-base-1556304 (accessed August 3, 2016).

Ryan, S. E. (2014), *Garments of Paradise: Wearable Discourse in the Digital Age*, Cambridge, MA: MIT Press.

Sagan, C. (1994), *Pale Blue Dot*, New York: Ballantine.

Salmoiraghi, A., and D. Akin (2012), "Review of Wearable Robotic Assistive Devices for Integration with Pressure Suit Arms," paper presented at 42nd International Conference on Environmental Systems (ICES), San Diego, July 15–19. https://arc.aiaa.org/doi/abs/10.2514/6.2012-3487 (accessed July 16, 2017).

Sanad, R., T. Cassidy, and T. L. V. Cheung (2012), "Fabric and Garment Drape Measurement—Part 1," *Journal of Fiber Bioengineering and Informatics*, vol. 5, no. 4, pp. 341–358.

Savage, A. (2015), "Adam Savage Incognito at Comic-Con 2015 (with Astronaut Chris Hadfield)," *Tested.* http://www.tested.com/art/

makers/534134-adam-savage-incognito-comic-con-2015-astronaut-chris-hadfield/ (accessed July 18, 2015).
Saville, B. P. (1999), *Physical Testing of Fabrics*, Cambridge, UK: Woodhead.
Schmidt, P. B. (2001), “An Investigation of Space Suit Mobility with Applications to EVA Operations,” PhD thesis, Massachusetts Institute of Technology. http://hdl.handle.net/1721.1/8105 (accessed July 16, 2017).
Schneiderman, D., and A. Griffith Winton (eds.) (2016), *Textile Technology and Design: From Interior Space to Outer Space*, London: Bloomsbury.
Schoeser, M. (2003), *World Textiles: A Concise History*, London: Thames & Hudson.
Science Museum (2016), “Couture in Orbit,” Event Programme.https://www.asi.it/sites/default/files/attach/notizia/couture-in-orbit-programme.pdf (accessed August 7, 2018).
Scmitt, P. B. (2001), “An Investigation of Space Suit Mobility with Applications to EVA Operations,” PhD thesis, Massachusetts Institute of Technology, Dept. of Aeronautics and Astronautics. https://dspace.mit.edu/handle/1721.1/8105#files-area (accessed December 19, 2016).
Seedhouse, E. (2011), *Trailblazing Medicine: Sustaining Explorers during Interplanetary Missions*, New York: Springer.
Seligman, L. (2016), “SpaceX’s Reusable Rocket: The ‘Holy Grail’ of Space Flight?” *Defense News*, January 6. http://www.defensenews.com/story/defense/air-space/space/2016/01/06/spacexs-reusable-rocket-holy-grail-space-flight/78319782/ (accessed January 7, 2016).
Seymour, S. (2008), *Fashionable Technology: The Intersection of Design, Fashion, Science, Technology*, New York: Springer.
Shaw, D. B. (2004), “Bodies out of This World: The Space Suit as Cultural Icon,” *Science as Culture*, vol. 13, no. 1, pp. 123–144.
Shaw, D. B. (2008), *Technoculture: The Key Concepts*, Oxford: Berg.
Shedroff, N., and C. Noessel (2012), *Make It So: Interaction Design Lessons from Science Fiction*, New York: Rosenfeld.
Siceloff, S. (2017a), “A New Market Emerges: NASA Partnerships Open the Path from Ground to Space,” NASA, March 27. https://www.nasa.gov/feature/a-new-market-emerges (accessed May 27, 2017).
Siceloff, S. (2017b), “New Spacesuit Unveiled for Starliner Astronauts,” NASA. https://www.nasa.gov/feature/new-spacesuit-unveiled-for-starliner-astronauts (accessed May 27, 2017).
Smith, A. (2005), *Moondust: In Search of the Men Who Fell to Earth*, London: Bloomsbury.
Smith, C. (2014), “Have Spacesuit, Will Travel,” TEDx Talk presented at TEDx Brussels, March 6. https://www.youtube.com/watch?v=bMlL0aV75VY (accessed May 27, 2017).
Solomon, S. (2016), “The Martians Are Coming—and They’re Human: How Settling Mars Could Create a New Human Species,” *Nautilus*, issue 41. http://nautil.us/issue/41/selection/the-martians-are-comingand-theyre-human (accessed August 7, 2018).
Southern, T. (2011), “Why Spacesuit Gloves Hold the Key to the Future of Humanity,” TEDx talk presented at TEDx Midtown, New York, May 9. https://www.youtube.com/watch?v=XxJOtPPC02k (accessed May 27, 2017).
SpaceX (2016), “Making History,” SpaceX, http://www.spacex.com/about (accessed August 1, 2016).
Sparke, P. (1993), *The Plastics Age: From Bakelite to Beanbags and Beyond*, Woodstock, NY: Overlook Press.

Stecker, P. (1996), *The Fashion Design Manual*, New York: Palgrave Macmillan.
Steimle, H., and C. Norberg (2013), "Astronaut Selection and Training," in E. Norberg (ed.), *Human Spaceflight and Exploration*, New York: Springer-Praxis, pp. 255–294.
Steyerl, H. (2011) "In Free Fall: A Thought Experiment on Vertical Perspective," *E-Flux Journal*, no. 24. http://www.e-flux.com/journal/24/67860/in-free-fall-a-thought-experiment-on-vertical-perspective/ (accessed June 22, 2017).
Stoppa, M., and A. Chiolerio (2014), "Wearable Electronics and Smart Textiles: A Critical Review," *Sensors*, vol. 14, pp. 11957–11992.
Strickland, P. (2012), "Do Space Hotels Differ from Hotels on Earth?" *Journal of Hospitality Marketing and Management*, vol. 21, no. 8, pp. 897–908.
Stylios, G. (2015), "Editorial: Spacesuits for Commercial Travellers," *International Journal of Clothing Science and Technology*, vol. 27, no. 2. https://doi.org/10.1108/IJCST-02-2015-0024.
Sun, H., X. Fu, S. Xie, Y. Jiang, and H. Peng (2016), "Electrochemical Capacitors with High Output Voltages That Mimic Electric Eels," *Advanced Materials*, vol. 28, no. 10, pp. 2070–2076.
Szewcyzk, S. (2015), "Exploring the Future of Fashion with Final Frontier Design," *Tech Times*, July 31. http://www.techtimes.com/articles/73515/20150731/final-frontier-designs-next-generation-space-suits.htm (accessed July 9, 2016).
Takahashi, H. (2003), "How do Plants Grow in Microgravity?" *Japan Aerospace Exploration Agency.* http://global.jaxa.jp/article/special/kibo/takahashi_e.html (accessed August 7, 2016).
Taras, Nick (2014), "Masturbating Chimpanzees, Boiling Blood and Spacesuits: The Human Body in Space," *Lifted Brow*. http://theliftedbrow.com/post/102388933162/masturbating-chimpanzees-boiling-blood-and (accessed May 12, 2015).
Tate, K. (2013), "Columbia Space Shuttle Disaster Explained," Space.com. http://www.space.com/19526-columbia-shuttle-disaster-explained-infographic.html (accessed August 30, 2016).
Telegraph (2014), "Models Dressed in Gucci Strut down Vertical Catwalk in Japan," *The Telegraph* [online], May 7. http://www.telegraph.co.uk/news/newstopics/howaboutthat/10813918/Models-dressed-in-Gucci-strut-down-vertical-catwalk-in-Japan.html (accessed July 17, 2016).
Thomas, K. S., and H. J. McMann (2012), *U.S. Spacesuits*, 2nd edn, New York: Springer.
Thorpe, A. M. (2009), *The Commercial Space Station: Methods and Markets*, Bloomington, IN: Authorhouse.
Timmins, M. (2010), "From Space Suits to Space Couture: A New Aesthetic," in Ulrike Landfester, Nina-Louisa Remuss, Kai-Uwe Schrogl, and Jean-Claude Worms (eds.) *Humans in Outer Space—Interdisciplinary Perspectives*, Vienna: Springer, pp. 183–203.
Tortora, P. G. (2015), *Dress, Fashion and Technology: From Prehistory to the Present*, London: Bloomsbury.
Tribbe, M. D. (2014), *No Requiem for the Space Age: The Apollo Moon Landings and American Culture*, Oxford: Oxford University Press
UCR Artsblock (2016), "Cosmokinetial Kabinet Noordung—Postgravity Art," *Free Enterprise*. http://sites.artsblock.ucr.edu/free-enterprise/cosmokinetial-kabinet-noordung-postgravityart/ (accessed August 3, 2016).
UK Department of Transport (2017), *UK Draft Spaceflight Bill*, presented to Parliament by the Secretary of State for Transport by command of Her Majesty. https://www.gov.

uk/government/uploads/system/uploads/attachment_data/file/592928/draft-space-flight-bill-web.pdf (accessed June 22, 2017).

Van Dusen, M. (2012), "Questions for Nicholas de Monchaux, Author of Spacesuit: Fashioning Apollo," *Txchnologist*. http://txchnologist.com/post/25510809371/questions-for-nicholas-de-monchaux-author-of (accessed January 5, 2018).

van Herpen, I. (2014), "About: Biopiracy," Iris van Herpen. http://www.irisvanherpen.com/about (accessed August 10, 2016).

Veblen, T. (1899), *The Theory of the Leisure Class: An Economic Study of Institutions*, reproduced online at http://www.gutenberg.org/files/833/833-h/833-h.htm (accessed July 29, 2016).

Vogler, A. (2005), "Design Study for an Astronaut's Workstation," SAE Technical Paper 2005-01-3050. http://papers.sae.org/2005-01-3050/ (accessed February 9, 2017).

Vogue (1905), "Paris," *Vogue*, April 27, pp. vi–vii.

Vogue (1913), "Fashion: The Gradual Approach of the New Silhouette," *Vogue*, April 1, p. 44.

Vogue (1917), "Fashion: Not Even Draperies Can Lure the New Silhouette from Its Straight and Narrow Way," *Vogue*, October 1, p. 67.

Vogue (1921), "Fashion: The New Silhouettes Bring Back the Corset and Demand Well-Fitted Brassieres," *Vogue*, October 1, pp. 77; 100.

Vogue (1930), "The 1930 Game of Twenty Questions," *Vogue*, October 27, pp. 45; 108; 110.

Vogue (1947), "In a Manner of Wearing, or How to Wear the New Silhouette," *Vogue*, September 15, pp. 141–142; 208.

Vogue (1957), "Cover Look," October 1, *Vogue*, p. 3.

Vogue (1960), "The Brief Hat, Weightless Dress," *Vogue*, February 15, pp. 82–83.

Vogue (1960), "The New Thread of the Weightless Matter," *Vogue*, March 15, pp. 136–139.

Vogue (1996), "The Straight and Narrow," *Vogue*, September, p. 578.

Vogue (2008), "Flights of Fancy," *Vogue*, September 1, pp. 684–695.

Wachhorst, W. (2000), *The Dream of Spaceflight: Essays on the Near Edge of Infinity*, New York: Basic.

Wakata, K. (2014), "Koichi Wakata Weekly Activity Report," vol. 19, January 25, 2014. https://www.youtube.com/watch?v=MlqXsl7kauo (accessed August 14, 2015).

Waldie, J. M., and D. J. Newman (2011), "A Gravity Loading Countermeasure Skinsuit," *Acta Astronautica*, vol. 68, no. 7–8, pp. 722–730.

Wall, M. (2013), "Ticket Price for Private Spaceflights on Virgin Galactic's SpaceShipTwo Going Up," Space.com, April 30. https://www.space.com/20886-virgin-galactic-spaceshiptwo-ticket-prices.html (accessed November 9, 2017).

Wall, M. (2017), "Boeing Unveils New Spacesuits for Starliner Astronaut Taxi," Space.com, January 25. http://www.space.com/35456-boeing-unveils-starliner-spacesuits-photos.html (accessed May 27, 2017).

Wang, Y., D. Wu, M. Zhao, and J. Li (2014), "Evaluation on an Ergonomic Design of Functional Clothing for Wheelchair Users," *Applied Ergonomics*, vol. 45, no. 3, pp. 550–555.

Watkins, S. M., and L. E. Dunne (2015), *Functional Clothing Design*, London: Bloomsbury.

Watts, N. (2011), *Around the World in 60 Minutes*, BBC Four, first broadcast 9 p.m. Monday, March 14.

Webber, D. (2010), "Point-to-Point Sub-orbital Space Tourism: Some Initial Considerations," *Acta Astronautica*, vol. 66, nos. 11–12, pp. 1645–1651.

Webber, D. (2017), *No Bucks, No Buck Rogers: The Business of Commercial Space*, Norwich: Curtis Press.

Weed, S. (2007), "High Dive," *Popular Science*, July, pp. 52–57.

Weislogel, M. M., D. R. Pettit, A. P. Wollman, and R. M. Jenson (2016), *Capillary Beverage Cup*, US Patent 20160088959 A1. http://www.google.com/patents/US20160088959 (accessed July 4, 2016).

Weitering, H. (2017), "Chanel 'Launches' Rocket at Paris Fashion Week," Space.com, March 7. http://www.space.com/35950-chanel-launches-rocket-paris-fashion-week.html (accessed March 10, 2017).

Westfall, Gary (2012), *The Spacesuit Film: A History, 1918–1969*, London: McFarland.

Whitmore M., K. McGuire, S. Margerum, S. Thompson, C. Allen, C. Bowen, B. Adelstein, S. Schuh, V. Byrne, and D. Wong (2013), "Risk of Incompatible Vehicle/Habitat Design," Human Research Program Space Human Factors and Habitability Element, NASA. https://humanresearchroadmap.nasa.gov/Evidence/reports/HAB.pdf (accessed August 18, 2016).

Winder, J. (2015), "Rubber Suction Pants Could Save Astronauts' Vision," Sen.com, http://sen.com/news/rubber-suction-space-pants-could-improve-astronauts-vision (accessed January 7, 2016).

Wood, A. (2014), "*Gravity* by Alfonso Cuaron (Review)," *Science Fiction Film and Television*, vol. 7, no. 3, pp. 441–444.

Woods, A. R. (1993), "The 'Cosmic Dancer': Sculpture and the Absence of Gravity," *Leonardo*, vol. 26, no. 4, pp. 297–301.

Woods, A. R. (1994), "Art In Space: The Spaceflight of the Cosmic Dancer Sculpture," *Earth Space Review*, vol. 3, no. 2. http://www.cosmicdancer.com/publications/Earth-Space-Review-1994.pdf (accessed August 7, 2016).

Woods, B. (2009), "A Political History of NASA's Space Shuttle: The Development Years, 1972–1982," in D. Bell and M. Parker (eds.), *Space Travel & Culture: From Apollo to Space Tourism*, Malden, MA: Blackwell, pp. 25–46.

Yuhas, A. (2017), "SpaceX to Send Two People around the Moon Who Paid for a 2018 Private Mission," *The Guardian*, February 28. https://www.theguardian.com/science/2017/feb/27/spacex-moon-private-mission-2018-elon-musk (accessed July 22, 2017).

Yotka, S. (2017), "Fashion Week Show Notes: Chanel Aims for the Stars," *Vogue*, March 7. http://www.vogue.com/article/chanel-fall-2017-paris-fashion-week-ready-to-wear-news-recap-karl-lagerfeld (accessed March 10, 2017).

Young, A. (2009), *Spacesuits: The Smithsonian National Air and Space Museum Collection*, Brooklyn, NY: Powerhouse Books.

Zakharine, D. (2013), "Preface: Electrified Voices: Medial, Socio-Historical and Cultural Aspects of Voice Transfer," in Dmitri Zakharine and Nils Meise (eds.), *Electrified Voices: Medial, Socio-Historical and Cultural Aspects of Voice Transfer*, Göttingen: Vandenhoeck & Ruprecht, pp. 11–30.

Zhang, L., J. Li, and Y. Wang (2013), "Development and Mobility Evaluation of the Female IVA Clothing," *Advanced Material Research*, vol. 821–822, pp. 677–680.

INDEX

2001: A Space Odyssey (film and book) 121

Adidas 5, 80–2
aerial (performance) 58, 60–1
see also vertical catwalk shows
aerial (view) 62, 64–7, 110, 118, 123
air travel 19, 21, 42
anchoring 13, 97–100, 104
Apollo program 13, 16, 24, 27, 83, 95
art *see* creative arts
astroculture 13–18
augmentation 100, 119

balance *see* stability
Barbarella 17–18
body shape 89, 100–2, 123
Boeing 2, 22, 76, 77–8
Boeing Blue (spacesuit) 77–8
boots *see* footwear
Branding 22, 37, 69, 76–7, 80, 81, 87, 91
Branson, Richard 11, 22
also see Virgin Galactic

CAD (Computer Aided Design) 113–15
Cardin, Pierre 11, 15, 16, 17
catsuit 34, 81
catwalk 45–51, 55, 56–7, 58–61, 137
Challenger disaster 20
Chayalan, Hussein 43–4, 137
cocktail glass *see* Zero Gravity Cocktail Glass
color 25, 28, 34, 38, 63, 77, 81
Columbia disaster 20, 122
commercial spaceflight *see* private sector
contact (between garment and body) 45–6, 50, 51–3, 55, 109–13, 116, 139
corsetry 19
Cosmic Dancer (sculpture) 27–8, 38
Cosmic Lifestyle Corporation 3, 32–3, 91
costume 10, 17–18, 73, 82, 109, 139
COTS (Commercial off-the-shelf) 85, 87–91, 105, 139
Courrèges, Andre 11, 15–16, 62–3
Couture in Orbit 37–9
creative arts 3, 11, 27, 27, 28, 29–31
creative industries 25 *see also* creative arts
crowdfunding 76
customization 5, 10, 69, 76, 87

David Clark Company 19, 76, 78
defamiliarization 11, 29, 32–3, 34, 35, 65, 95–6
design 8, 11, 13–17, 31–3, 34, 36, 38, 39, 41, 80–3, 89, 90–1, 100, 122, 127–8
see also product design
doffing 71–3, 107, 124–5
donning 34, 71–3, 124–5
drape 6, 8, 38, 44, 45–7, 50–5, 56–61, 94, 107–16
drapemeter 109–10
dressing *see* donning
Dubois, Kitsou 29–30, 96–7

eating and drinking 32–3, 56, 137
elastic 51, 73, 89, 111, 112
elevation view 61–7, 110, 123–124
EVA (extravehicular activity) 4, 5, 7, 13–14, 78–9, 94–5
exclusivity 6–7, 16, 33–4, 75, 76, 80 *see also* luxury

falling 28, 41, 44–5, 97, 104, 109–10, 115, 121–2
final frontier (the) 13

Final Frontier Design 5, 75–6, 119, 139
fit 34, 81, 88, 89, 101, 102, 104, 106–7, 111–12, 123
flammability 84, 86, 87, 122, 126
flight suits 69, 80–83 *see also* pressure suits
footwear 15, 16, 20, 25, 26, 63, 73, 78, 97–100, 119, 121–2, 139
form-fitting 11, 34, 35, 38, 52, 81, 110, 126

gloves 5, 70, 73, 75, 95, 100
Gravity Loading countermeasures 52, 101, 102–4, 127

Hadfield, Chris 4, 73, 94
hands 56–57, 97–100, 104 *see also* gloves
helmets 13, 14, 15, 16, 17, 18, 20, 25, 95, 119, 129
hem 16, 44–5, 47, 51–5, 58, 60, 63–4, 66–7, 73, 100, 104, 106, 107, 109, 110–13, 124, 138
hero 4, 11, 13, 75, 82, 128, 129
Herpen, Iris van 45, 47–9
horizon 26, 27, 62, 63–4
Hyper Space Couture Contest 36–7, 38, 39

Icarus 58, 121–2

Jet Age 42, 81

Kennedy, John F. 11–13, 24

leisure 3, 5, 78, 126–128 *see also* SpaceX
length (of garment) 45, 52–5, 102, 124, 129, 138
lint 86, 121
luxury 3, 6, 33–4, 128

Mars 2, 4, 8, 9, 23–4, 25, 73, 78, 84, 89, 115, 118, 122, 128, 130
mass 85–6, 87, 89–90, 126
Matsui, Eri 37–8
McQueen, Alexander 45–7, 48
microgravity *see* weightlessness
mobility 52, 70–1, 75, 78–9, 95–6, 101, 103, 129
modularity 71, 73, 81, 107, 124
Moss, Kate 45–7, 48
movement 5–6, 29–30, 44, 46–7, 52, 54–6, 66, 71–3, 81, 93–100, 103–4, 121, 138
Musk, Elon 2, 3, 5, 11, 22, 23–4, 70, 82–83, 119, 128, 136

Neutral Buoyancy Lab 52
neutral posture 6, 8, 50, 93, 99, 104, 106–7, 113, 118

off-the-shelf *see* COTS
Orbital Outfitters 76, 82
orientation 8, 28, 29–30, 56–68, 99, 104
Orndoff, Evelyne 86, 89–90

parabolic flight 28–31, 35, 37–8, 39, 80, 88, 95, 97, 99, 108–9, 115–16, 120, 128
pattern cutting 64–6, 106–7, 113–14
post-gravity 11, 29, 94, 120–1
posthumanism 119–21
posture 5–6, 8, 10, 37, 50, 73, 88–9, 97–100, 105–7, 113, 115–16, 118, 123
pressure suits 5, 7, 19, 70, 75, 78–79, 83–4, 101, 103, 133
private sector 1–3, 21–5, 69–70, 79–81
product design 32–3, 35–6, 121, 122
protection *see* safety
public engagement 74–6

Rabanne, Paco 15, 17–18
reduced gravity aircraft *see* parabolic flight
rigged displays 50–1, 57
Roberts, Julian 64–6
Rocketplane Global 36, 136
routine wear 34, 83–9, 104, 106, 122
runway *see* catwalk

safety 1, 4, 5, 7, 20, 69–70, 72, 3, 78–80, 81, 84, 87, 118, 125, 126
Sagan, Carl 27, 119
science-fiction 5, 11, 13, 16–18, 25, 39, 49, 57, 73, 82–3, 108–9, 121–2, 129, 130
sculpture 31, 38, 67 *see also* creative arts
sewing 19, 130
Sharman, Helen 55–6

shoes *see* footwear
silhouette 8, 11, 14, 34, 35, 36, 37, 45, 52–3, 55, 58, 61–2, 63–4, 67, 81, 100, 107, 116, 123–4, 126, 129 *see also* drape
sizing 101–102, 113–114, 140
Space Age 6, 11–21, 25, 34, 36, 38–9, 42, 62–3, 81, 129
space tourism *see* tourism
Space Wear Vision 89–90
spacesuit 4–5, 7, 13, 16, 18, 19, 20, 21, 36, 69–79, 82–4, 94–6, 119, 120, 125, 129, 130
spacewalk *see* EVA
SpaceX 1–3, 22–4, 69, 80, 82–3, 91, 128, 135, 136
stability 8, 44, 56, 73, 96–100, 102–4, 121, 124, 125 *see also* contact
Star Trek 17
style 3, 6–7, 13–16, 36–7, 39, 69–70, 73–5, 77–8, 79–83, 116, 127, 130
subtraction cutting *see* Roberts, Julian

tailoring 78, 106–7
testing 28, 55, 88, 109, 128–9
textiles 37, 39, 42–3, 86, 90, 108, 114–15, 125–6
tourism 2, 3, 5–9, 11, 15, 21, 22, 25, 33, 34–5, 69, 78, 82, 117, 126, 128, 129

underwater 41, 52–6
undressing *see* doffing
Upside Down (film) 121–2
USSR *see* Russia

V2 Variable Vector Countermeasure Suit 121
vertical catwalk shows 41, 58, 60–1
verticality *see* orientation
VEST 6, 89–90, 112
Viktor and Rolf 45, 47, 57–8, 59
Virgin Galactic 1, 2, 3, 5, 22, 23, 69, 70, 80–2, 83, 89, 91, 115
virtual simulation *see* CAD
Vogue 8, 42–3, 45, 46, 61, 137

walking 93, 100, 103–4, 115
water *see* underwater
wearable technology 37, 70, 74, 103–4, 118–22, 139
wedding dress 37–8, 55
weight (of fabric) 42, 86–8, 108, 126
weightlessness 3, 9, 11, 28–30, 31, 33, 34–6, 39, 41, 43–5, 47, 49, 50, 55, 56–7, 72–3, 93, 94–5, 121, 127
  appeal to tourists 3, 7, 28, 34–5, 117
  depictions of 44–5, 51 *see also* Upside Down (film) and 2001: A Space Odyssey (film and book)
  designing for 3, 5–8, 11, 39, 67, 100, 116, 118, 122, 123–6, 127, 129
  effects on clothes 1, 5–6, 36, 38, 50, 52, 56, 88, 108, 109, 112–15
  effects on body shape 89, 93, 100–2, 104, 123 *see also* neutral posture
  effects on fabric 86, 108, *see also* drape
  experiences of 11, 27–8, 30, 35, 41, 56, 57, 94, 95–7, 104, 107, 108, 115, 120, 121 *see also* defamiliarization
  in fashion history 16, 18, 42–3
  performance in 29–30, 31
  sewing in 130
  virtual simulation of 114–15
wellbeing 6, 84, 89

Y-3 *see* Yamamoto, Yohji
Yamamoto, Yohji 5, 80–2

Z-2 (spacesuit) 4–5, 73–5, 81, 130
Zero Gravity Cocktail Glass 3, 32–3
Zero Gravity Corporation 29, 37, 80
zero-G flight *see* parabolic flight